A NATURAL HISTORY OF LAND'S END

Other Books by Jean Lawman

A Natural History of the Lizard Peninsula
A Naturalist's Year – Wildlife at Land's End
Skokholm, an Island Remembered

A NATURAL HISTORY OF LAND'S END

Jean Lawman

TABB HOUSE
Padstow

First Published 2002
Tabb House, 7 Church Street, Padstow
Cornwall, PL28 8BG

Copyright © Jean Lawman 2002

ISBN : Please see back cover

The right of Jean Lawman to be identified as the author of this work has been asserted by her estate in accordance with the Copyright, Designs and Patents Act, 1998.

All rights reserved, except within the terms of licence of the Copyright Licensing Agency Ltd and the Publishers Licensing Society Ltd.

British Library Cataloguing-in-Publication Data.
A Catalogue record of this title is available from the British Library.

This book is for all the people who appreciate and work to conserve our Cornish wildlife.

CONTENTS

	Page
FOREWORD by Stella Turk	ix
ACKNOWLEDGEMENTS	xi
LIST OF ILLUSTRATIONS	xii
INTRODUCTION	xix
1. ENVIRONMENTAL INFLUENCES	1
Geography	"
Geology	4
Weather	7
2. COASTLAND	10
Plants	15
Birds	28
Mammals	44
Basking Sharks, Sunfish and Turtles	48
Reptiles	50
Butterflies	51
3. MOORLAND	55
Plants	58
Birds	71
Mammals	"
Butterflies	72
4. WETLAND	74
Plants	"
Birds	85
Dragonflies	92

	Page
5. WOODLAND	96
Plants	97
Birds	108
Mammals	115
Butterflies	117
Dragonflies	118
6. FARMLAND	119
Plants	121
Birds	"
Butterflies	132
7. TOWNS AND VILLAGES	134
Plants	"
Birds	139
8. CONSERVATION	144
APPENDIX	148
Plants of the Land's End Peninsula	"
Birds	176
Butterflies	187
and Dragonflies	189
Mammals	190
REFERENCES	193
INDEX	195

FOREWORD

Jean Lawman's first book about the Land's End peninsula was written in the form of a diary and that book, with her own line-drawings, showed her to be a naturalist/artist, with the close and sensitive observation one would expect from such a rare person. This production, like the earlier one and the charming volume on the Lizard peninsula, is decorated with her drawings, with the addition of some photographs.

She is now giving readers another treat in her second book on the Land's End. It enables us to explore in detail with her the wildlife of the coast, moors, farms, marshes and ponds as well as the little stands of woodland. She also touches on the role played by the estates and the wildlife one can see in the towns and villages. Inevitably the Cornish hedge is present in most of these habitats – a habitat of its own, made by man and variously adorned by nature. The landscape was formed millions of years before humankind was conceived, and there is a clear exposition of the geological background on which the present panorama is staged. The weather of this windswept dynamic landscape is also discussed.

There are records galore – flowering plants and ferns, mammals, birds, reptiles, butterflies and dragonflies – each species described in its appropriate time and space setting, often with a note on the aesthetic delights provided, or some fragment of folklore or a local name, like *caseg coit* 'mare of the wood' for the Green Woodpecker, referring to it sounding like a neighing horse. Additionally she adds little gems from other naturalists who, like the plants and animals themselves, are native or introduced; thus pages are studded with choice morsels from books by other writers who have been entranced by this peninsula, including J. C. Tregarthen, W. H. Hudson, C. C. Vyvyan and J. T. Blight. As a dedicated member of Cornish Dolphin Watch, Seaquest Southwest and Coastwatch over many years, the author is skilled in the identification of members of the 'mega' fauna, including sharks, seals, turtles and sunfish and most of the cretaceans that occur in British waters. She lists those that can be sighted from time to time from her study areas and describes how Bottlenose Dolphins sometimes 'perform' near the Minack Theatre to the great delight of audiences!

There is a section on the various conservation measures that apply to this south westerly tip of Britain. Much of it is part of an Environmentally Sensitive Area within which farmers and other owners (including the National Trust and Cornwall Wildlife Trust) follow practices designed to enhance the wildlife. The reference list gives the author's sources, suggesting further reading. For those keen to know the range of species found, there are very useful and comprehensive appended lists.

There is little doubt but that more books, fiction and non-fiction, have been written about Cornwall than any other region of the British Isles, with the possible exception of London. Vast and diverse as the resultant library is, there is certainly room for this one which will be read and cherished by many readers, drawn to this part of Cornwall by such strange place names as Bostraze, Nanquidno, Chy-en-hal, Treryn Dinas and Kenidjack.

Jean Lawman states that the intention of her book is to arouse interest in the subject of natural history, to give a baseline of information and to promote the conservation of what is still a relatively unspoilt landscape. The aim of this foreword is to show that she does this, and much more.

Stella Turk

ACKNOWLEDGEMENTS

Many people have helped me with this work and particularly with the appendices. Among them I would like to thank especially Jon Brookes, John Worth, John Swann, Pat Sargeant, Ray Dennis, Stella Turk, and Katie Herbert from Penlee House Museum. Len Margetts provided me with much useful criticism in the early days as I tackled the seemingly never ending plant list.

LIST OF ILLUSTRATIONS

Line drawings   Page

Maps: The Land's End Peninsula	xv
Simplified Geology Map of the Peninsula	xvi
Topography Map of the Peninsula	xvii
Wild Carrot *Daucus carota ssp. carota*	xviii
Common Alder Alnus glutinosa	3
Three-cornered Garlic *Allium triquetrum*	8
Danish Scurvygrass *Cochlearia danica*	12
Spring Squill *Scilla verna*	"
Kidney Vetch *Anthyllis vulneraria*	19
Sea Campion *Silene uniflora*	"
Oxe-eye Daisy *Leucanthemum vulgare*	20
Thrift *Armeria maritima*	"
Bird's-foot-trefoil *Lotus corniculatus*	21
Rock Samphire *Crithmum maritimum*	23
Sea Beet *Beta maritima*	24
Hottentot Fig *Beta maritima*	27
Wheatear	30
Great Skua	31
Razorbill	33
Oystercatcher	38
Purple Sandpiper	39
Atlantic Grey Seal	46
Common Adder	50
Marsh Fritillary	53
European Gorse *Ulex europaeus*	58
Lousewort *Pedicularis sylvatica*	59
Red Campion *Silene dioica*	60
Ground Ivy *Glechoma hederacea*	61
Wall Pennywort *Umbilicus rupestris*	62
Devil's-bit Scabious *Succisa pratensis*	"
Himalayan Balsam *Impatiens glandulifera*	63

	Page
Merlin	65
Raven	"
Wren	69
Grayling	72
Grey Willow. Male Catkins *Salix cinerea ssp. oleifolia*	75
Grey Willow. Female Catkins *Salix cinerea ssp. oleifolia*	"
Marsh Violet *Viola palustris*	78
Cornish Moneywort *Sibthorpia europaea*	79
Hemlock Water-dropwort *Oenanthe crocata*	81
Ragged Robin *Lychnis flos-cuculi*	82
Meadowsweet *Filipendula ulmaria*	"
Bulrush *Typha latifolia*	84
Moorhen	87
Coot	88
Sedge Warbler	89
Emerald Damselfly *Lestes sponsa*	94
Buds: Sycamore *Acer pseudoplatanus*	98
Buds : Beech *Fagus sylvatica*	99
Haws: Hawthorn Fruit *Crataegus monogyna*	101
Hips: Fruit of the Dog Rose *Rosa canina*	"
Blackthorn *Prunus spinosa*	102
Hazel *Corylus avellana*	103
Common Elder *Sambucus nigra*	"
Bluebell *Hyacinthoides non-scripta*	105
Pink Purslane *Claytonia sibirica*	"
Lesser Celandine *Ranunculus ficaria*	"
Ramsons *Allium ursinum*	106
Dog's Mercury *Mercurialis perennis*	107
Great Tit	110
Great-spotted Woodpecker	111
Green Woodpecker	112
Yellow Bartsia *Parentucellia viscosa*	122
Field Pansy *Viola arvensis*	"
Greater Stitchwort *Stellaria holostea*	124
Cow Parsley *Anthriscus sylvestris*	"
Field Scabious *Knautia arvensis*	"

	Page
Golden Rod *Solidago virgaurea*	125
Yarrow *Achillea millefolium*	"
Chamomile *Chamaemelum nobile*	126
Yellowhammer	128
Swallow	129
Alexanders *Smyrnium olusatrum*	135
Red Valerian *Centranthus ruber*	137
Red Admiral	138
Small Tortoiseshell	"
Bullfinch	141
Starling	143
Corncockle *Agrostemma githago*	150
Traveller's Joy *Clematis vitalba*	155
Common Toadflax *Linaria vulgaris*	163
Ribwort Plantain *Plantago lanceolata*	167
Musk Mallow *Malva moschata*	203

Colour photographs   Between pages 76 & 77

The contrast of slate and granite: the harsh outline of Mylor Slate at Carn Gloose
Smooth contours of the granite cliffs at Gwennap Head
Old field systems at Zennor
Sea mist on the exposed cliffs at Pendeen showing the carpet of *Thrift Armeria maritima* and Sea Campion *Silene uniflora*
Wild flowers on a stone hedge on south coast cliffs
Wind-pruned hawthorn on exposed hedge
A Celtic cross with Three-cornered Garlic *Allium triquetrum* growing around base, in a sheltered situation
Beautiful Demoiselle *Calopteryx virgo*
General distribution: English Stonecrop *Sedum anglicum*
Oxeye daisies *Leucanthemun vulgare*
Wetland: Cotton-grass *Eriophorum angustifolium*
Farmland: Corn Marigold *Chrysanthemum segetum*
Spring in Lamorna Woods with bluebells *Hyacinthoides non-scripta*, ferns and young Sycamore leaves

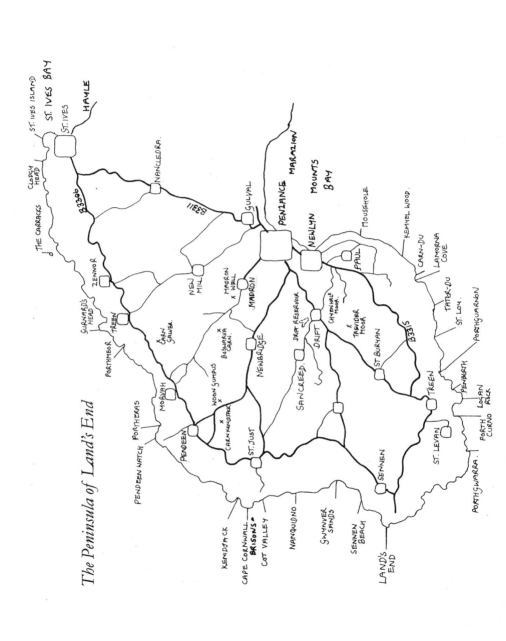

Simplified Geology Map of the Peninsula

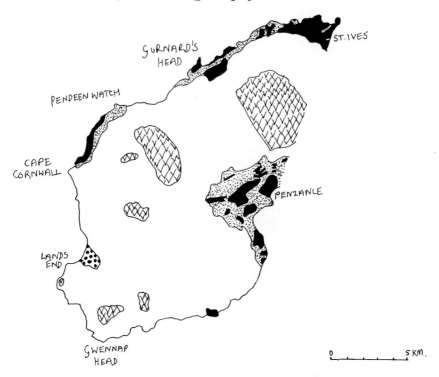

KEY

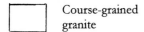 Course-grained granite

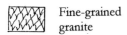 Fine-grained granite

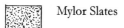 Mylor Slates

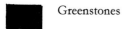 Greenstones

 Blown Sand

Topography Map of the Peninsula

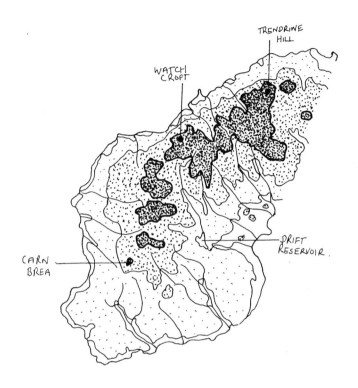

HEIGHT IN FEET (METRES)

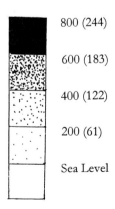

800 (244)

600 (183)

400 (122)

200 (61)

Sea Level

Wild Carrot
Daucus carota ssp. carota

1
ENVIRONMENTAL INFLUENCES ON THE PLANTS AND WILDLIFE

Here in the west we seldom get a taste of real winter on the autumn side of Christmas. Sometimes indeed we get no taste of it at all, and geraniums, nasturtiums and gazanias will 'winter throu' as we say.

C.C.Vyvyan (1952)

While the main theme of the book is natural history, it is important to consider, in broad terms, some aspects of the geography, geology and weather.

Geography

The peninsula lies in the extreme south-western part of the British Isles, and this book describes the area to the west of the B3311 road that connects Penzance with St Ives.

Most of the land is made up of an outcrop of granite which determines its general shape and form. It is bounded by a wild and rugged coastline, much of which consists of cliffs between 50m and 90m high, broken by rocky coves and sandy beaches. Of the latter, Treen, Porth Curno, Porth Chapel, Sennen and Gwenver are the most popular. The sand, on close examination, is seen to consist wholly or partly of tiny shell fragments; this was observed by Blight at Porth Curno: 'the beach here is entirely of sand composed of comminuted shells of the most delicate and beautiful structure, and is left so smooth by the waves that it is almost a pity to make footprints there'. Small sand dune systems behind the beach add variety to the flora at Sennen and Gwenver.

In places, headlands composed of harder, more weather-resistant rocks protrude dramatically into the Atlantic Ocean, the two most spectacular being Gurnard's Head and Cape Cornwall. The cliffs are highest in the north and west and, in places, rise sheer from the sea to an extensive erosional plateau around 130m, which is itself broken by a series of granite tors running in a south-westerly direction from St Ives across to St Just. Like many parts of Cornwall, the peninsula has been subjected to fluctuating sea levels and, inland of the plateau, a degraded cliff-line is ascribed to high sea levels in the Tertiary Period. This is thought to be contemporary with the plateau surface of the Lizard peninsula. If this was the case, the land above 130m would have formed a series of islands cut off from the rest of Cornwall by a channel between Hayle and Penzance.

From the highest of these tors – Watch Croft (252m), or from other more accessible ones such as Carn Galver or Zennor Hill, there are excellent views across the plateau to the cliffs and sea beyond. The action of snow and ice in glacial times, beginning just over 1.6 million years ago, resulted in erosion of the tors and a great scattering of boulders across the plateau. These have been used to build the intricate network of stone walls that enclose the irregularly-shaped fields. The foundations of many of these walls date back to the Bronze Age (2400-800 BC). Long ago, people were unable to explain this phenomenon of large boulders spread across the land, hence the old Cornish legends telling of angry giants who hurled these granite monsters from the craggy tops. The stony nature of the landscape is reflected in everything from cottages and farmhouses to walls, stiles and bridges, a natural harmony of man with the landscape. This has great appeal to artists and naturalists. Most modern buildings look obtrusive and detract from the character of this fascinating and ancient terrain.

The granite moors are drained by streams running in a north-westerly or a south-easterly direction along natural fault lines. Those flowing to the north coast follow a shorter and steeper course than those flowing south across more gently graded and cultivated countryside.

To follow the course of one of these streams, flowing from the moors to the south coast, illustrates the varying countryside. The

Newlyn River, for example, rises near the Men Scryfa (an ancient inscribed stone) on the open moors, traverses farmland, is dammed to form a reservoir and then, before reaching the coast in urbanized Newlyn, meanders through mixed woodland. Its picturesque course, which is a little changed, was described by Blight in 1861: 'It is bordered sometimes by masses of purple heath and sweetly-scented flowers; it splashes over mill-wheels too, and forms pretty waterfalls over moss-covered rocks; it is crossed by other rustic bridges but none so picturesque as this one at Buryas, which is partly clad with ivy, and shaded by ash trees. The stream now glides down the fertile and wooded valley of Trereife'. The old bridge at Buryas has been replaced but it is well depicted in a painting by Stanhope Forbes and also in old photographs. The building of the dam at Drift to form the reservoir in 1969 is the most notable change.

Remains of a forest, which was submerged by high sea levels after the last glaciation, can be seen at very low tides between Wherry Rocks and Lariggan Rocks, especially when storms have removed quantities of sand. This was part of an extensive forest beneath what is now Mounts Bay. Tree-stumps rooted in a peat layer 1.7m thick are sometimes exposed, and from the remains oak, hazel, birch and alder trees have been identified.

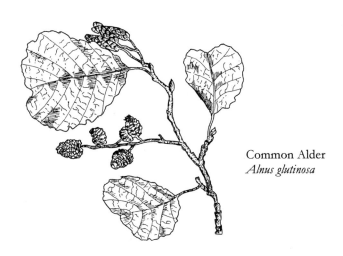

Common Alder
Alnus glutinosa

Geology

The theory of plate tectonics revolutionized the understanding of geology. The earth's crust is now known to consist of a number of plates which move against each other or pull apart, powered by convection currents generated by the hot molten interior of the earth. In this way, new crust is formed as molten rock wells up and solidifies in the gaps, or one piece of crust is forced beneath another as the plates converge. In very simple terms, Cornwall is on the trailing edge of a continent carried by one of these plates. For our purposes we are concerned with a more precise account of the local geology.

A large elongate mass of granite called a batholith underlies most of Devon and Cornwall and it tends to form high ground because it is generally harder than the surrounding rock. An offshoot of this granite mass forms the bulk of the Land's End peninsula, dominating all but an area in the south-east around Penzance and strips on the north coast. It is an igneous rock, that is, it arose from the solidification of molten magma issuing from beneath the earth's crust. This was intruded beneath the older cover rocks during the early Carboniferous/Permian Period (280 million years ago) and during the gradual cooling process it acquired its crystalline structure. Granite rock varies a great deal between and within districts. Weathered granite is grey and usually covered in lichens, but when it is freshly broken it is seen to be made up of large crystals of white or translucent grey quartz, white or pink feldspar and shiny black and white micas. The size of the crystals varies, depending on the rate of cooling. Colour, alignment and association with other minerals are other factors which affect its general appearance. Around the coast, most particularly the stretch between Porthgwarra and Land's End, wind, rain and salt spray, have weathered the granite into fantastic shapes. In thick fog the rocks loom larger than life and their strange and unworldly shapes create an eerie atmosphere. Little imagination is required to detect all kinds of grotesque animals, giants and such-like sculpted out of the rocks. Horizontal joints may be weathered so much as to form piles of granite, and sometimes one rock becomes precariously balanced on top of another, giving rise to the 'logan rock' phenomenon, best illustrated at Logan Rock itself, located near Treen. Hollowed-out, water-filled rock basins are another feature of weathering. They form when slightly acid rainwater lies in a shallow dip, weakening the feldspars and loosening the quartz grains, resulting in further erosion of the basin.

The cover rocks into which the granites intruded were Mylor Slates of late Devonian age (about 400 million years ago). These originated from sedimentary rocks: shales, clay, mudstones or sandstones that were laid down as sediments usually on the sea bed and then altered by heat and pressure to form slate, which is then classed as metamorphic rock. These, as cover rocks, were then weathered to expose the granite. The Mylor slate formation forms part of the broken band of older rocks around the coastal fringes and can be seen in several places, e.g. Porthmeor and Pendeen. The rock varies in colour between green-grey, blue-grey, dark grey or black.

Also exposed within this band, and contemporary with the Mylor slates, are the greenstones and pillow lavas. The former are a group of igneous, intrusive rocks that are more resistant to weathering than granite and which help to form headlands like Gurnard's Head and St Ives Island. Greenstone was once quarried extensively at Penlee Quarry, between Newlyn and Mousehole, and made a good quality roadstone. In the field these rocks are dark with a sharper outline that contrasts with the softer, more rounded granite rock. This is seen very clearly on the headland at Tater-du where the dark greenstone is splashed with light green, tufted lichens and also brilliant yellow Xanthoria. Here too, lower down near the lighthouse, veins of pink-coloured granite can be seen penetrating the greenstone.

The other rock found in these exposures are pillow lavas, formed when volcanoes emptied beneath the sea and the lava rolled down slopes and then solidified in pillowy forms while retaining cavities of gas and steam. These are clearly seen on the uppermost tip of Gurnard's Head.

Two other igneous rocks were formed during the consolidation of the granite as it intruded into the Mylor slates. These are aplite and pegmatite, veins of which can be seen cutting through the older formation in Porthmeor Cove.

Also in association with the formation of the granite were gases and heated liquids given off by the molten magma, causing metallic minerals to be deposited, like tin and copper. These gave rise to the mining industries, once such an important part of the overall economy in Cornwall. Further alteration of some small areas of granite by hydrothermal activity produced kaolin or china clay which is still mined at Bostraze, near St Just.

The glaciations took place in the Pleistocene Age (beginning 1.64 million years ago), and followed a period of submergence five million years ago when the 130m marine platform was formed. It is generally accepted that there were four main glaciations, each lasting several thousand years with warmer periods in between, although it is now considered to have been more complex than that. During the greatest glaciation, the Wolstonian, sea ice lay off the north Cornwall coast and some of it may have passed over the Isles of Scilly. The land was then subject to periglacial conditions, i.e. subject to alternate freezing and thawing, with a climate similar to that of Greenland today. This led to an assorted mass of rock debris moving down slopes and valleys by a process known as solifluction. These 'head' deposits are exposed on the coast in many places and are seen as an orange-brown layer of crumbly, mud-like material at the top of some of the lower cliffs, usually where valleys intersect the coast, e.g. at Cot Valley and St Loy Cove.

Changing sea levels were a feature of the glaciations, and the advance of the sea ver the land in interglacial times left a series of planed off surfaces known as raised beaches. They are commonly seen embedded beneath the head deposits on the coast, often as a horizontal line of rounded stones at a height of about 5-8m. There are good examples at Nanquidno and Cot Valley and above them in Cot Valley, a layer of angular stones represent debris formed in freeze-thaw conditions when glaciers were active in the north of Britain. A 20m raised beach was once exposed at Penlee Quarry and water-worn pebbles have been found at this height at Morrab Place in Penzance. There is evidence of another at about 30m.

For practical geology guidance one should refer to two colour-illustrated Holiday Geology guides by Tony Goode (see list of references at the end of the book).

The soils of the district, being mostly derived from granite, tend to be poor, acid and therefore relatively infertile. The natural vegetation

ENVIRONMENTAL INFLUENCES... 7

cover of the district is heathland, with scrub and mixed woodland where there is some degree of shelter. The most fertile areas are in the river valleys which contain alluvial soil. In contrast to the north, much of the southern part of the peninsula is cultivated with heavy use of fertilizer and additional liming.

Weather

In the early part of the twentieth century W. H. Hudson, the celebrated naturalist and writer, spent some time exploring this area, making acute observations on animals and plants, man included. During this period he recorded certain events occurring in the month of December, which struck him as unusual because of their unseasonal character. They included Peacock butterflies flying on sunny days, the song of the Corn Bunting (now extinct in the district) and, lingering at Land's End, a Wheatear 'which had decided that it was not necessary to fly all the way to North Africa to find a place to winter in'. Daffodils *Narcissus cultivars* and Lesser Celandines *Ranunculus ficaria* flowering on Christmas Eve come as no surprise to the people of Cornwall. A Comma butterfly seen on Christmas Day in 1991 was a typical but rather untimely event.

Such occurrences, together with a host of others, are simply indications of the relatively mild temperatures experienced in this extreme south-west corner of Britain. They are due to the ameliorating effect of the land being almost surrounded by ocean and, to a large extent, the warming influence of the Gulf Stream. This brings with it much moisture, so that rainfall is higher than in the east of Britain. Coastal fogs are common, occurring when warm, moist air from the Atlantic meets colder air masses over the land, triggering, one hopes in this age of automation, one or all of the lighthouse foghorns into action. These latter are strategically placed at Tater-du on the south coast, the Longships off Land's End and Pendeen on the north coast, all now unmanned. Detailed information about Cornwall's lighthouses can be found in 'Cornwall's Lighthouse Heritage' by Michael Tarrant. The average daily temperature for this region is 10.9°C as compared with 9.5°C for Blackpool, 260 miles further north. Moreover, the temperature range is less extreme, with warmer winters and cooler summers. The highest temperatures occur in July and August, but the driest weather is usually between April and June, when spring droughts are occasionally prolonged into summer. It is during these months that the peninsula is at its most beautiful with the freshness of spring lingering in the air and a proliferation of flowers on the cliffs and in the hedgerows.

This geniality of climate is also responsible for the untimely and sporadic flowering of many wild plants throughout the winter months, particularly on the more sheltered south coast. Blooms of Sheepsbit *Jasione montana*, Sea Campion *Silene uniflora*, Thrift *Armeria maritima*, Kidney-vetch *Anthyllis vulneraria* and Betony *Stachys officinalis* are commonly seen well out of the normal flowering season and in February 1989, I recorded a Foxglove *Digitalis purpurea* in full flower. The ubiquitous Red Campion *Silene dioica*, which is extremely common in Cornwall, has been found flowering in every month of the year as has Common Gorse *Ulex europaeus*, and in the very mild February of 1989, two months earlier than usual, this plant was in full bloom.

The general lack of hard frosts means that gardeners can grow plants from the Mediterranean and sub-tropical areas quite successfully, and these aliens can become naturalised quite easily as has the Hottentot Fig *Carpobrotus edulis* from South Africa and Three-cornered Garlic *Allium triquetrum* from the Mediterranean. There are, however, exceptional winters such as 1986-'87 when temperatures fall as low as -11°C and naturally many plants and animals suffer as a result. The Hottentot Fig grows well except for those plants which face a north-east aspect; they succumb to bitter cold winds from this direction. Wind-clipped trees and bushes add greatly to the character of the landscape, reflecting a rather more severe aspect of the weather, i.e. the prevalence of strong winds. Land's End experiences an average of thirty gale days a year compared with London which has less than five and Blackpool which has ten. Except in very sheltered places, plants are

Three-cornered Garlic
Allium triquetrum

constantly and mercilessly battered by these salt laden winds, many of them only surviving because they have developed special growth forms or certain other adaptations to the harsh conditions. The salt winds burn and dry out the buds on the windward side of the trees and bushes, killing them off, while those on the leeward side survive and grow to give the familiar wind-clipped profiles.

A proper appreciation of the landscape, especially the dramatic coastline, is possible only when it has been experienced in all weathers, particularly the storms which carve and shape the rock and render so much of the land inhospitable but wildly beautiful. Blight makes this very point when writing of the Land's End cliffs: 'To be appreciated the Land's End cliffs must be seen in calm and in storm, in sunshine and in cloud. Walk on the turf fragrant with wild flowers, sit among the sea pinks and follow with the eye the numerous birds pursuing their vocations, watch the waves as they "play the summer hours away" and the vessels as they creep along near the land – for the sky is fair and the sun bright. How fearful is the change when the blasts howl and shriek around the cairns, and the deafening roar of billows fills the air.'

The above account was intended to be a background for those who wish to study the natural history in more detail. The following chapters describe the different habitats to be found within the peninsula and some of the plants and animals likely to be encountered in them.

Visitors should equip themselves with the 1:25,000 Ordnance Survey Map of the Land's End Peninsula, 'Explorer 7'. On this map public footpaths are explicitly marked. They traverse the countryside from farm to farm, from church to village, etc. and can be used in conjunction with the coastal path and those traversing the moor to plan interesting and varied circular walks.

Please respect the countryside – enjoy the flowers without picking them and the wildlife without causing unnecessary disturbance.

2

COASTLAND

The great rampart of cliffs that holds back the Atlantic is broken here and there by beaches of white sand, or minute shells, or by coves, into which fall trout streams that rise in the granite hills above.

J. C. Tregarthen (1904)

The coast path that skirts the peninsula from Penzance round to St Ives passes through infinitely variable scenery. The nature of the coastline, being indented with bays and intersected by valleys, means that the path follows a circuitous route with a lot of uphill and downhill walking. The landscape also changes according to degree of exposure and the extent of man's activities like mining and farming.

From Penzance heading westwards the cliffs are gently sloping, fairly overgrown and broken occasionally by partly wooded valleys, carved by small rivers or streams that may trickle unnoticed beneath piles of boulders to the sea, as at Lamorna and St Loy. Some of the smaller streams become obscured in marshy gullies that are grown over by tall, rank vegetation before they spill down low cliffs onto the rocks below. Penberth and Porthguarnon are exceptionally pretty coves, the former a working cove owned by the National Trust, where the small fishing boats are winched up onto the granite slipway. The latter is a steep, rocky valley that is less accessible. More popular is Lamorna Cove with its attractive harbour and the fine sandy beaches of Porth Curno and Porth Chapel.

Further west, when the small cove at Porthgwarra has been left behind, the coast is completely open to the full force of storm-ridden seas and high winds. The granite cliffs are sheer and spectacular with fine examples of pinnacles, stacks and arches and, instead of cultivated fields, they are backed by a wild expanse of rolling heather and gorse moor. This weather-embracing landscape includes many lofty headlands, the most spectacular of which is the Land's End itself, whereafter the path continues round to the old coastguard hut at Sennen (now a National Trust visitor centre), and onto a much gentler section incorporating the long sandy beaches of Sennen and Gwenver with their small dune systems.

Beyond this stretch the rocky valley bottom at Nanquidno marks the beginning of a section extending round to the north coast, where tin mining was once the main occupation. There are plenty of reminders of this historically fascinating period such as, for example, the derelict engine houses with their tall chimneys and stonework draped in creepers and ferns, now home to Jackdaws and Kestrels. The adits themselves can be seen bored into the valley sides and cliffs, but it is not safe to explore them. Kenidjack, Botallack and Portheras have particularly fine examples of these ruins. Around Pendeen, which was the main mining area, some of the old buildings, far from being austere and depressing, are architecturally splendid with the red, brown and grey colours of the stonework blending perfectly. The National Trust have sensitively restored some of their buildings and Levant mine is open to the public, as is Geevor mine, run by the Pendeen Community Heritage Group. Nearby, Pendeen Lighthouse is managed by the Trevithick Trust and is open for viewing.

After Pendeen, continuing on to St Ives, the coast is wild and rugged with hardly a tree in sight, only scattered farmhouses with their higgledy-piggledy stone walls and the high granite tors beyond. Stunted Blackthorn *Prunus spinosa*, the occasional clump of willow *Salix sps*, and leggy gorse are the only tall vegetation that survive well in this wildest and least inhabited northern region. It is open and bleak indeed when it is compared with the south coast, so one feels that it really belongs to the wheeling seabirds, the seals and the small birds that eke out a living here.

One great advantage of the varying aspect of the coastline is that in inclement weather there is always a degree of shelter to be found on one side or the other, with a cove or a valley somewhere that is out of the wind altogether. On occasion, though, it is worth braving the full force of the gale to watch the enormous, white-streaked waves roll in from the boiling ocean and to witness the power behind them as they crash against the cliffs in great plumes of snowy white spray. It is an awesome sight and a humbling experience.

The sunny aspect and greater degree of shelter found on the south coast are reflected in the earlier and more luxuriant growth of wild flowers here. In spring when the slopes between Mousehole and Porthgwarra are strewn with daffodils, celandines, scurvygrass *Cochlearia danica* and *C. oficinalis* and other early species, the north coast will be relatively bare save for a few Dog Violets *Viola riviniana* flowering beneath dead Bracken *Pteridium aquilinum* and odd clumps of Primroses *Primula vulgaris* in damp clefts and gullies. Generally Thrift, Sea Campion Spring, Squill *Scilla verna*, and the late flowering heathers are

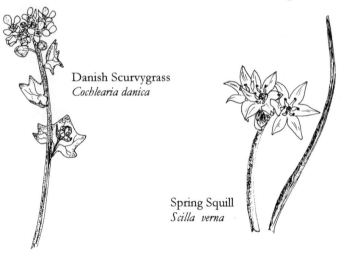

Danish Scurvygrass
Cochlearia danica

Spring Squill
Scilla verna

common all around the coast, but they put on a better show along the northern stretch. There is at least two weeks' difference in the flowering times of the north and south coasts, but while it may not be so early or many of the flowers quite so abundant, the north coast has some of the finest scenery in Cornwall.

The difference in exposure is one of the reasons why the land has been more extensively cultivated in the south and there are usually arable or grass fields directly inland of the coast path. Not so many years ago there were hundreds of tiny, terraced fields on the cliff slopes where the local people grew flowers and early potatoes. These charming little fields, enclosed by primitive stone walls or planted hedges, are described by many authors in the late nineteenth and early twentieth centuries. The earliest potatoes in England were grown at Tater-du according to Folliot-Stokes. Today these fields are overgrown and neglected but there is a certain magic in the air when, as early as February, sprays of yellow daffodils and white narcissi burst through the dead bracken and brighten up the cliffs on the dullest of days. When picnicking by these miniature fields, one wonders whether the hard working flower-pickers of the past had the time to appreciate the beauty of the flowers, set against a cobalt sea. Along the south and west coasts, both on the cliff slopes and in the valleys, these small, stone-walled enclosures are still in evidence in lots of places even though many have become overgrown. Daffodils still grow in some of them, especially around St Loy and Tater-du. Pink Gladioli or 'Jacks' were sometimes grown, and they are often seen in association with these old fields or by the coast path. They flower in May and appear most years around the headland to the east of Penberth.

Also in the valley at Penberth, Sweet Violets, Kaffir Lilies, and a variety of vegetables are grown in small fields that are enclosed by tall Escallonia and privet hedges. Many of the fields had been neglected for years but recently much work has been put in to revive this traditional industry. Both the abandoned and the cultivated fields, together with their hedges, are especially good for butterflies and small birds, as well as being examples of a simple and less intensive type of agriculture. The National Trust owns much of Penberth and deserves credit for its sensitive upkeep of the valley and the cove.

Early potatoes are still grown on the cliff-slopes above Scathe Cove, east of Penberth, and occasionally at St Loy. They are always the first to be lifted.

The following quote from Folliot-Stokes (1908) describes the small fields at Tater-du: 'We came across as curious a collection of little gardens as could be found anywhere. Scores of little plots, each one surrounded by an elder hedge ten to fifteen feet high. So thick are these hedges and so small the plots they enclose, that the whole thing seems a cunningly constructed maze, through which you wander from one little garden to another, till a sense of direction vanishes. After the primeval tangle of the cliffs, and the weedy spaciousness of the contiguous farm fields, the shaded tidiness of these well-tilled enclosures is strange indeed.'

The north coast has never been cultivated to the same extent because of its inhospitable nature and because of the difficulty of removing the stones that are spread across the land. In the past sheep and cattle were put out on the cliffs to graze. Today this would be non-profit making and, apart from grazing by the National Trust for conservation purposes, rabbits are the only animals that nibble the turf. The Trust owns a flock of rare Manx Loghtan sheep which are kept on Bosigran Cliff to graze the headland and its associated field systems. On other land that they own they have made special arrangements with local farmers to put out cattle at specified times, for example on Treveal Cliff. Grazing is considered by most conservationists to be a good management technique, benefiting some of the plants and animals, but it is a subject open to many arguments, as is the whole concept of purposeful management. Past removal of grazing stock has led to a general loss of short, herb-rich grassland, with many of the small, attractive flowering plants being unable to compete against coarser grasses. Many are disappearing as taller, rank growth takes over. There have been changes in the bird populations too, with an increase in smaller birds that nest in the scrub, like the Whitethroat and Dunnock, and a corresponding drop in numbers of other species that need short, grassy areas in which to feed. Most significant in this respect is the now extinct Chough, or Cornish Daw as it used to be known, and another is the Wheatear.

Stone walls, built precariously low down on the cliff-edge, are reminders of the days when grazing on the cliffs was common practice. They were there to stop animals straying over the cliff and one wonders, in this age of high technology and heavy machinery, at the amount of work involved in constructing these lovely old walls way down beneath the rambling coast path. Further evidence of grazing is provided by many historical references to the animals themselves or to features like sheep paths.

Before tourism came to Cornwall in a big way, many folk in the peninsula made their living from fishing and farming, often both, while others worked in the mines. Mining brought changes which affected farming as it depleted the labour force on the land, and it also physically altered the appearance of some of the valleys and cliff-faces. Most, but not all, of the associated buildings are on or near the coast and many are in a derelict state now but, together with the old shafts, they provide an alternative habitat for wildlife. Birds nest in holes in the stonework, ferns like the Black Spleenwort *Asplenium adiantum-nigrum* colonise the damp and shady entrances to the adits and bats live in the underground shafts.

Plants

Small areas of woodland are able to grow in some of the coastal valleys and sometimes right down by the sea, as at St Loy and Lamorna. The combination of trees and water is always enchanting and it was on this theme that Blight wrote a short piece about St Loy, which could easily have been written today: 'The trees extend to the verge of the cliff, a strange combination of luxuriant foliage with wild and savage rocks against which the waves are ever beating'. In the wooded section of the coast path to the east of the cove, some of the trees were coppiced long ago. This is a traditional form of management which ensures a constant supply of fuel while, at the same time, supporting a wealth of wildlife. These particular trees are planted and are mostly Turkey Oak, Sycamore and Sweet Chestnut, and they have the most wonderful wind-clipped profiles, standing as they do on a low cliff above the beach. Another special feature at the top of the western side of the beach is a band of dwarf Sessile Oak pushing up through the boulders to a height of only three or four feet. Close by is a tangled mass of Wild Hop *Humulus lupulus*.

For many reasons, not least the planting of foreign species of tree, the general character and composition of the woods have changed and this is discussed under the woodland section of the book. It is also interesting to note that some of the valleys, like Penberth and Kenidjack, were once extensively wooded, as shown by old photographs, but now little remains. Conversely, there are prints of Lamorna showing that at least the bottom part of the valley was once more open.

Halfway between Mousehole and Lamorna there is a block of woodland known as Kemyel Wood that consisted mostly of large, coniferous Monterey Cypress and Monterey Pine trees with a few others like Flowering Cherry and Horse Chestnut. The wood was devastated in the storms of 1990 and some replanting has been carried out by the Cornwall Wildlife Trust which owns it. The wood has an interesting history because, in maturity, it resembled a plantation although it was never intended as such. Some of these enormous trees were originally planted as windbreaks around approximately one hundred tiny, terraced fields in which early flowers were grown. Others were planted as a shelter belt for the fields above. Remains of earth and stone walls can still be seen within the wood and there are women alive today who remember working in these fields, connected to the farm by a rough cart track. To suddenly find oneself in the middle of this once much gloomier wood after the sunlit, flower-strewn cliffs was a strange experience indeed. The ground in many places was dark and bare where the slowly rotting pine needles had acidified the soil and this, together with excessive shade, prohibited the growth of most plants with the exclusion of ivy and an occasional bluebell *Hyacinthoides non-scripta*. The wood was once a silent place with no flush of insects to attract small birds, a situation typical of most conifer woods. Now, however, the storms have made huge clearings and it supports more wildlife. Fuchsia has now spread over a large area on the edge of the wood.

Limited by the degree of exposure and by shallow soil, rank scrubby growth has developed on many of the cliff tops in the south and, to a lesser degree, in sheltered parts of the other coasts. This is more so now that cultivation and grazing have virtually ceased. The coast path winds through a tangled mass of gorse, blackthorn, bracken and bramble *Rubus fruticosus*, crossing wet gullies where elder *Sambucus nigra* and willow grow. Now and again it passes through open areas that are lightly grazed by rabbits, but the majority of wild flowers grow down the slopes where the undergrowth and rabbits are more restricted as the exposure is greater.

Where there are old field systems, particularly just to the west of Mousehole, there may be hedges of elm, privet or Escallonia, the latter an evergreen shrub from Chile with bright pink flowers and shiny serrated leaves. This and some other introduced shrubs like Euonymus and Pittosporum species are widely planted on the Isles of Scilly because of their ability to form thick hedges and because they are very tolerant of salt-laden winds. Before these shrubs were introduced, Tamarisk *Tamarix gallica* was often planted as a windbreak. Originating from the Mediterranean, it is not nearly so effective in making a thick hedge, but it still survives in many places, for example at Sennen, Porth Curno and Porthgwarra. The informal growth and attractive sprays of green, feathery foliage with their tiny, fragrant, pink flowers suit our rugged coast.

Blackthorn comes into blossom before the leaves break out on its damson coloured branches. Then it is as if a light cover of snow has fallen on the dark bushes. Some years are better than others, but in a good year there is no more lovely sight than a mixture of flowering blackthorn and gorse, with a blue sea for a background. Hudson revelled in the beauty of the flowering gorse: 'With a clear blue sky beyond I do not know in all nature a spectacle to excel it in beauty. It is beautiful perhaps because the blossoming furze is not the 'sheet of gold' it is often described, but gold of a flame-like brilliance sprinkled on a background of darkest, harshest green.'

The gorse that turns valley, moors and cliffs such an intense yellow in spring is the taller and more widespread European Gorse. Later in the year, simultaneously with the heathers, the low-growing Western Gorse *Ulex gallii* comes into flower and the combination of the two makes a wonderful mosaic of bright pink and yellow. One of the best places to see this is on the open heath beyond Porthgwarra, although places on the north coast, particularly around Zennor, are certainly equal to it. Unfortunately, a severe heath fire destroyed a most beautiful stretch of heather and gorse between Porthgwarra and Nanjizal in 1995. The flowers of Western Gorse are a deeper orange-yellow and this species is not so widespread as the former, being confined to heathland only.

Climbing plants grow entangled among the coastal scrub, their sprawling and often sticky stems finding the necessary support and degree of protection among the taller plants. These include Honeysuckle *Lonicera periclymenum*, Wild Madder *Rubra peregrina*, Cleavers (Goosegrass) *Galium aparine*, Hedge Bedstraw *Galium mollugo* and Hedge Bindweed *Calystegia sepium*.

When the daffodils have died away many a cliff slope takes on a blue wash as thousands of bluebells come into flower. Instead of trees, they use bracken as their nurse. So, just as April is the month for gorse, May is the month for bluebells. Afterwards, tender, young bracken fronds, with their tips curled to protect the shoot, take over to form a dense cover of deep green.

Bluebells are not the only flowers that are associated with woodland to be found on the open cliffs. Another is the Wood-sorrel *Oxalis acetosella*, but this is to be found mostly on the north coast and then only very locally. It has a delicate, nodding flower with mauve veins etched on the white petals and pale green, notched leaves divided into three lobes. It also grows beneath bracken.

Dense undergrowth may make it difficult to venture off the path onto the headlands and lower slopes, but those that are accessible are well worth exploring because it is here that some of our most beautiful plants grow. These plants are especially adapted to withstand constant physical battering and excessively salty conditions which dry them out. There are several ways that plants can protect themselves and conserve water; some adopt a cushion-like growth, others have fleshy leaves or a waxy cuticle, and many of those not confined to the coast are found to be more than usually hairy.

May and June are generally considered to be the best month for flowers. Apart from the Bluebells, Thrift, Sea Campion and Kidney Vetch put on a fine display. Pink, white and yellow blooms make a patchwork of colour around the rocks and Spring Squill, with flowers like clusters of little mauve stars, carpets bare and trampled areas. This latter was described by Hudson as 'a little drop of cerulean colour in a stony desolate place'. There are many places, especially headlands, that could be recommended for a visit, but there is nothing like the sudden and unexpected pleasure of discovering for oneself these natural rock gardens.

Cyclical changes in weather patterns cause variations in the abundance of these flowers from year to year. Prolonged spring or summer droughts that kill off the competitive grasses are usually followed by a prolific flowering season, and I remember a breathtaking sight at Pendeen with Thrift, Sea Campion and Spring Squill flowers so dense that hardly a green leaf was visible between them. This was on the cliffs here in 1991 after two summer droughts.

Wild Carrot *Daucus carota ssp. carota*, ancestor of the Garden Carrot,

Kidney Vetch
Anthyllis vulneraria

Sea Campion
Silene uniflora

is one of the most attractive of our umbelliferous plants. It is common on the cliffs here and, while it looks lovely in the summer with its delicate foliage and powdery, whitish-pink umbels, it is equally so in the winter when the skeleton of the flowerhead appears in dark silhouette against a bright sea. A unique feature of this plant is that some of the umbels have a tiny, deep crimson flower in their centre. It blooms in June and July, usually in grassy places on the cliff tops and down the slopes.

Also in June, many of the cliffs are bright with Oxeye Daisies *Leucanthemum vulgare* and they are particularly abundant around the Logan Rock/Porth Curno area where they mix with foxgloves and button-like, mauve Sheepsbit *Jasione montana*. The spectacle of all these flowers strewn along the cliff edge, with the classically beautiful Pednevounder beach below and the rugged headland of Treryn Dinas in the distance, must be one of the loveliest in Cornwall.

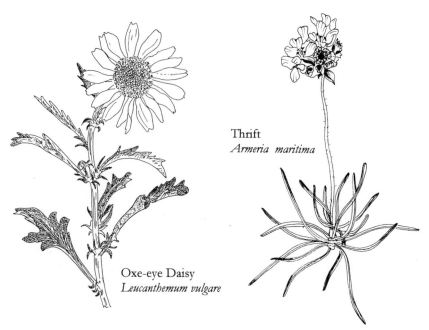

Thrift
Armeria maritima

Oxe-eye Daisy
Leucanthemum vulgare

Rocky crevices, walls and soils laid bare by fire are suitable places to look for the pale pink flowers of the English Stonecrop *Sedum anglicum*. The small, rounded, fleshy leaves, usually tinged red, are characteristic of the stonecrop family and are unmistakable. This plant, together with the purplish-pink flowers of Wild Thyme *Thymus polytrichus* and Bird's-foot-trefoil *Lotus corniculatus*, which may be brilliant yellow, orange or red, make extravagant splashes of colour on the granite in summer. The flecked grey, crystalline structure of the latter sets off the flowers beautifully. Rocky outcrops and Cornish hedges are ideal habitat for these plants, but they can also be seen embellishing some of our ancient monuments like the walls of the Ballowall chambered cairn on the coast near St Just.

Occasionally fires break out on the cliffs, as happened at Porthgwarra, Treen and Kenidjack in the hot, dry summer of 1995. They can inflict serious damage on the plant communities, but for several years after the burn, opportunists like foxgloves, red campion and bird's-foot-trefoil come up and put on a fine display. It is not always a bad thing for very overgrown areas to be opened up but, at the wrong time of

COASTLAND

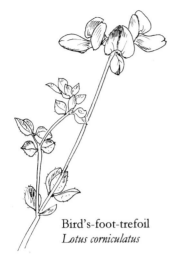

Bird's-foot-trefoil
Lotus corniculatus

year, an extensive burn can be detrimental to plants as well as birds, reptiles and invertebrates, and can have very serious consequences for those latter that are sedentary and have a limited distribution.

In early June the first flowers on the Bell Heather *Erica cinerea* begin to appear, but it will still be six weeks or so before it reaches the peak of its flowering and deep rose red patches appear on the heath above the clifftop. The inflated, bell-shaped flowers cluster around the stem as do the whorls of needle-like, bright green leaves. In the dying process the flowers turn a tawny red and as this happens the Common Heather or Ling *Calluna vulgaris* breaks out into one-sided sprays of lilac pink flowers. Sometimes the feathery, olive green or brown leaves of this plant spread over granite stone in a most attractive manner, seen to best effect on the north coast. One other heather is common here but is confined to damp soils. This is the Cross-leaved Heath *Erica tetralix*, clearly quite different from the others in appearance, with its compact, one-sided clusters of inflated, pale pink flowers and grey-green, hairy leaves in whorls of four that give it the 'cross' in its name. Its singular beauty inspired the Rev. C.A. Johns to write: 'The part of the flower nearest the stem is of a lighter colour than that which is exposed, where it deepens to a delicate blush, the whole flower appearing as if modelled in wax'. A pure white variety of this species crops up occasionally.

The Cornish Heath *Erica vagans*, which is so plentiful on the Lizard, does not occur on the peninsula apart from one small patch near Porth Chapel that was found in 1990, the origin of which is uncertain. Similarly Dorset Heath *Erica ciliaris* is found in one place only, on Chyenhal Moor near Drift, where it was originally planted in 1934 and still exists. The hybrid Erica watsonii *Erica ciliaris* × *Erica tetralix* was recorded here before 1980.

In early summer, irregular, red patches appear on the heather and gorse, which on close inspection are seen to be made up of myriads of thread-like stems heaped together. This is Common Dodder *Cuscuta epithymum*, a parasitic plant which produces globular bunches of tiny, white flowers in summer.

On the coastal heathland the orchid family is represented by hundreds of spikes of the Heath-spotted Orchid *Dactylorhiza maculata* ssp. *ericetorum*, bearing flowers which vary in colour from dark pink to white. The leaves are usually, but not necessarily, dark-spotted. The exposed heathlands of Porthgwarra and the north coast are excellent locations for this species.

In the short turf between heather tussocks, several other plants are found that are easily identifiable when they are in flower. Common and Heath Milkwort *Polygala vulgaris* and *P. serpyllifolia* have small, but very brightly coloured, flowers of deep pink or blue and earned their name because they were supposed to increase the milk yield in cattle. On the other hand, Lousewort *Pedicularis sylvatica* was not well thought of because it was supposed to cause louse infestations in these same animals. It is a short, squat plant, but the tubular flowers are a pale rose-pink. More people were familiar with common plants in the past because so many of them were used for medicine or in local craftsmanship. Some were used as natural dyes, for example a yellow dye was extracted from Dyer's Greenweed *Genista tinctoria*, a prostrate plant that is related to gorse. The incredible brilliance of its yellow flowers is enhanced by the background of shiny, dark green leaves on which they lie. It is locally common on the north and west coast in grassy areas on the cliffs or on the edge of heathland, and is prolific on the north side of Cape Cornwall.

In a few sites on the lower cliff slopes between Treryn Dinas (an iron age cliff castle) and Land's End, two very special plants grow, and they are usually found together because they require the same conditions, i.e. south facing, dry rocky slopes with very little soil. They complement each other beautifully; the one with delicate sprays of pale mauve flower and the other with clusters of bright yellow, daisy-like blooms. They are Rock Sea Lavender *Limonium loganicum*, which has now been classified as a species in its own right, and Golden-samphire *Inula crithmoides*. These are very localised and sometimes grow in precarious situations, but are well worth looking out for. I once watched dozens of Painted Lady, Red Admiral and Tortoiseshell butterflies visiting a colony of these plants on a steeply sloping cliff; it was an unforgettable sight.

Scrambling among the rocks on a calm day with the sea gently lapping below is a pleasant enough pastime and it is only when a huge swell is rolling in, lashing surf and spray against the lower cliff faces, that one experiences the true hostility of the environment. Even so, some plants have adapted to living in these extreme conditions by being extremely tough and fleshy. They are able to retain water to such an extent that they can withstand the desiccating effect of excessive salt, while their physical toughness enables them to survive periodic battering by gale-force winds. One of these hardy individuals is Rock Samphire *Crithmum maritimum*, a perennial, with greyish-green, pinnate

Rock Samphire
Crithmum maritimum

leaves and umbels of tiny, yellow flowers. The leaves of this plant were once collected, pickled and then sold as a delicacy in the London markets. It is delicious when prepared this way, but it can also be simply chopped and added to salads. The flavour is slightly mustard-like. The Rev. C. A. Johns describes the preserve: 'The younger leaves if gathered in May, make, when sprinkled with salt and preserved in vinegar, one of the best pickles'. Another edible plant of the lower cliff zone is Sea Beet *Beta maritima*, which is very robust with tall, green flower spikes and leaves that look and taste like spinach. It is cooked in a similar way, but it is important to pick the young and tender leaves otherwise it is very bitter.

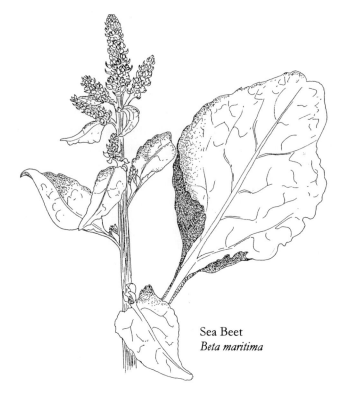

Sea Beet
Beta maritima

In certain other exposed situations, particularly where there is a good supply of nitrogen from seabird guano, biennial Tree Mallow *Lavatera arborea* grows up tall and stately with striking, pink flowers that are borne in clusters up the woody stem. There is a network of purple veins radiating from the dark central part of the petals. Fine specimens of this magnificent plant occur at Tater-du and Cape Cornwall. They can grow up to ten feet tall and at that height they begin to resemble small trees.

Common Scurvygrass is a fleshy plant that is very common in damp places, particularly beside streams that trickle down gullies to the sea.

A strong, sweet perfume is given off by the white, four-petalled flowers in spring. It has an earlier flowering relative, the Danish Scurvygrass *Cochlearia danica*, which is smaller and has a mauve tinge to the white flowers. This one will grow in drier, more exposed places, for example on the south-facing slopes of Kenidjack Cove and Cot Valley. In common with many other cliff plants, it is much more abundant in years that follow a summer drought. The name scurvygrass comes from the general belief that it was once eaten by shipwrecked sailors to prevent scurvy because it is rich in vitamin C.

Wet flushes arise from water percolating down the cliff slopes either as small streams or as run-off from the tops. This distinct habitat supports a variety of tall, attractive plants like Purple-loostrife *Lythrum salicaria*, Hemp-agrimony *Eupatorium cannabinum*, Common and Water Figwort *Scropularia nodosa* and *S. auriculata*, Greater Bird's-foot-trefoil *Lotus pedunculatus* and Fleabane *Pulicaria dysenterica*. Most have flamboyant, pink or yellow flowers, but the tiny figwort flowers are fascinating because their beauty lies in their burgundy-red colour and the unusual two-lipped structure.

In a few places on the south coast there are small stands of Common Reed *Phragmites communis*, the plant which is so extensive on nearby Marazion Marsh. It is really more beautiful in its mature, dried state when stem and leaves are faded buff and the dark brown, feathery plumes glint silver in the late autumn sun. Royal Fern *Osmunda regalis* looks best in mid-June when the newly grown fronds shine luminous green in the early morning light, or later on in the autumn when it turns deep russet and is one of the richest of the autumn colours. The places to look for this most handsome of ferns are streamsides, wet flushes and valley bottoms, especially by the coast. Porthgwarra and Porthguarnon have quantities of it and some specimens may reach six to eight foot in height.

Another fern, Sea Spleenwort *Asplenium marinum*, occupies damp and shady rock crevices, usually at a low level on the cliffs. It is very noticeable in that its shiny, bright green fronds with their rounded segments are usually the only vegetation growing in these dark cavities. It has a fairly regular growth form and looks wonderful when it grows in quantity at the entrances of some of the local caves. Blight was certainly impressed when he discovered it near Mousehole in a cave where it still flourishes today: 'At the entrance it is about fifty feet high, the roof is a mass of dark green, being entirely covered in ferns, which hang down in a most graceful manner; perhaps the Asplenium marinum; grows nowhere so fine as it does here'.

Some of the damp gullies on the coast are flushed with carpets of pale yellow primroses in April and May. This is more so on the exposed soils of the north coast where there is less competition from taller plants. They are very abundant on the slopes around Zennor, especially at Veor Cove.

The unstable nature of sand dunes, due to constantly shifting sand, results in a specialized flora developing. This changes composition from the seaward end to the more stabilized dune turf further inland. Dune plants are especially adapted to withstand being differentially covered with sand, having special vegetative mechanisms to enable them to spread like, for example, the extensive rhizome systems in the Sand Sedge *Carex arenaria* and Marram Grass *Ammophila arenaria*. They also have special features to give them physical protection such as the tightly inrolled leaves of the latter species. The sand dunes at Sennen and Gwenver are very small and have suffered greatly in the past from erosion by tourists. For this reason restoration schemes have been set up in the past by the Penwith District Council and the County Council to try to restore the dunes and to prevent further damage. Large areas were fenced off and various methods were tried to trap sand, which were partially successful, and this was followed by replanting of Marram Grass. The old fences are still there. However, the interesting plants can be seen growing outside the fences as well as within them. The grey-green, prickly leaves of the Sea Holly *Eryngium maritimum* are most noticeable, but even more so are the lovely thistle-like, blue flowers which appear in summer. Equally unusual is the Sea Bindweed *Calystegia soldanella* with its large, trumpet-shaped flowers, striped boldly in pink and white. Pink, vetch-like Restharrow *Ononis repens* and yellow Ladies Bedstraw *Galium verum* are easily recognizable, but there are two smaller plants that require careful scrutiny. They each have tiny, pink flowers that look similar, one being the Common Storks-bill *Erodium cicutarium* and the other Dove's-foot Crane's-bill *Geranium molle*. The latter has pinnate leaves, rather than rounded and no notches on the silky petals. Both can be found in habitats other than sand dunes.

Still in this locality, but growing high up on the shore, are patches of Sea Sandwort *Honckenya peploides*, a small, fleshy plant which has tiny, cryptic flowers that are yellowish-green and an extensive rhizome system.

For the most part this account has dealt with plants that have their native origins in this country, but there are certain aliens which were brought to Britain as garden plants and, finding the climate congenial, managed to escape and spread rapidly at the expense of our native flora. The Hottentot Fig *Carpobrotus edulis* from South Africa grows rampantly on some cliff faces here, as it does on the Lizard peninsula where it poses a threat to some of the rare plants of that botanically famous area. It forms a thick mat of creeping stems from which arise long, fleshy, three-sided leaves and large, daisy-like flowers that are sometimes pink and sometimes yellow. Its sporadic appearance in places that are totally inaccessible is thought to be due to the activity of gulls pulling away pieces of stem for nest building. Its spread has been measured at 50 centimetres a year and it is nearly impossible to eradicate successfully since its sudden disappearance creates further problems of soil erosion. The only plants killed are those exposed to bitter north-east winds. It has one known attribute, as far as the naturalist is concerned, in that it provides a safe refuge for adders, particularly as it has fire resistant properties. It has taken over considerable areas at Penberth, Porthgwarra, Porth Chapel, Cot valley, Cape Cornwall and probably several other coastal sites which otherwise would be colonized by our own more attractive natives.

Hottentot Fig
Carpobrotus edulis

The other introduction that is sufficiently common to have a significant effect on our flora is the Three-cornered Garlic. Its flower is superficially like a white bluebell and often described as such by visitors who are not familiar with it, but in fact they are not related at all. Its main characteristics are the soft, three-angled leaves and the strong smell of garlic it gives off when crushed. Among other places, it has encroached onto the cliffs where it mixes and successfully competes with the bluebell. In some instances it has completely taken over from them. Being quite a pretty plant, it is generally well accepted, but not by conservationists who do not usually view introductions by man as part of the natural order of things.

Birds

The serious ornithologist will be well aware that the Land's End Peninsula is best known for its migrant birds and for its spectacular passage of seabirds in particular weather conditions. Its geographical position dictates that it is on the route for many millions of common migrants which make journeys north to their breeding grounds in spring, and south again in autumn to winter feeding quarters. Mixed in with them are a small number of rare and uncommon species, and also vagrants which have strayed off course from their normal migration path or overshot their destination because of adverse weather conditions or inexperience. Indeed, birds resident in America turn up every year in various places on the west coast of Britain during the late autumn months, having been driven across the Atlantic while undergoing their north-south migration along the eastern Atlantic seaboard. Most of them probably do not live very long, arriving in an exhausted state and finding themselves in an alien environment. One exception is a Black Duck on Tresco, Isles of Scilly, which stayed and bred with a Mallard, the result of which were a number of 'Blackards'. Another survivor, on Stithians reservoir, was a Pied-billed Grebe which won the affections of a Little Grebe and produced hybrid offspring. Also with its origins in North America, a Ring-necked Duck has been a regular winter visitor on Drift Reservoir.

Migrants generally follow coastal routes and so peninsulas and islands are convenient stopping-off places for feeding and resting or for sitting out bad weather. In the autumn this peninsula, the Lizard and the Isles of Scilly are invaded by a host of twitchers, together with an element of less fanatical birdwatchers, hopeful of locating rarities to add to their long lists of 'lifers' or simply to observe the migration of common species. The west coast valleys particularly are very productive, having limited cover for small birds like warblers and flycatchers, while fields and hedgerows are scanned for shrikes, pipits and the like. The search is thorough and precise, almost like a military operation, and consequently there is seldom a year which does not turn up something exciting. For example, in 1987, hundreds of birdwatchers arrived to see the Northern Parula and the Swainson's Thrush, and later on, in 1990, the Yellow-throated Vireo, which was a first record for Europe, all of which were located in valleys on the west coast. More recently, sightings of rare birds include a Veery (American), a Yellow-billed Cuckoo (American), a Blue Rock Thrush (southern European) and a Greenish Warbler (eastern European). Details of county records may be found in the excellent reports produced annually by the members of the Cornwall Birdwatching and Preservation Society.

Twitching has been popular for a long time now and some birdwatchers are only really satisfied with sightings of new species, while others find as much pleasure in witnessing a 'fall' of common migrants. A fall occurs on a night when clear, starlit skies are suddenly obscured by thick cloud or fog, grounding hundreds and thousands of migrating birds. The following morning previously deserted cliffs, bushes and trees are alive with birds. Warblers, finches, thrushes and many others are brought down and many thousands of birds may be involved, the species depending on the time of year since migration dates vary from one to another. Many ex-lighthouse keepers, themselves regrettably an extinct species in Britain, will have witnessed falls of this kind, when birds that have become disorientated in bad weather at night are attracted to the white flashing lights, dashing themselves against the glass and falling onto the balconies. Other birds simply become too exhausted to continue and perish by drowning in the sea. In the early 1900s W.E. Clarke wrote a fascinating account of his ornithological experiences on lighthouses all round the coast of Britain and France, which included an account of small birds being pursued by hawks around the lanterns.

Migration begins in March with early arrivals like Black Redstarts, Chiffchaffs and Wheatears. Among the followers are Willow Warblers, Whitethroats and Sedge Warblers during April and May. Some of them continue north and others stay to breed locally. Later migrants include Spotted Flycatchers and Swifts, which time their breeding to coincide with the peak of flying insects. After a quiet period in June and July the return migration gets underway in August, with Chiffchaffs, Sedge Warblers and Spotted Flycatchers on the move again, this time to the south. September, October and November sees the main autumn passage, with numbers inflated by the year's juveniles, and then lastly there is an influx of thrushes, finches and other birds from further north or from the continent to overwinter here. Starling flocks also increase, with numbers of immigrants from the continent. Late migrants include Wheatears and Black Redstarts. Dates of arrival and departures are remarkably constant for individuals, as ringing studies have shown. The above account is simplified and there are many other species seen regularly on migration, for example Whinchats, Ring Ouzels, Grasshopper Warblers and many more, as well as a variety of wading birds and birds of prey (raptors). Apart from those species which prefer open areas like Wheatears, small passerines usually find their way up the valleys where they find food and shelter in the cover vegetation.

Wheatear

Thousands of seabirds feed and migrate in the waters that surround Cornwall, i.e. the English Channel, the Irish Sea and out in the Atlantic Ocean itself. Strong west or south-westerly gales drive them inshore, often a long way up the Bristol Channel or onto the south coast of England. Then, if the winds veer to the north west and remain strong, the birds in the Bristol Channel are forced down onto the north Cornwall coast as they return on their westward passage out to the open sea. For this reason, St Ives Island is nationally famous for its seabird passage, both for numbers and variety. The birds cross the bay in their thousands, and concentrate as they round the island, giving spectacular views to birdwatchers positioned by the coastguard lookout. Most of the large movements take place between August and November, when gales are frequent, and they mostly consist of shearwaters, gulls, skuas, auks and gannets,

Great Skua

plus more unusual species like petrels, phalaropes and terns. It is a most impressive sight, especially to the novice and even for those who have seen it all before, and usually there are a number of rarities among them to keep the old-timers interested. A seawatch on October 7th 1988 was evidently most memorable: 74 Grey Phalaropes, 998 Pomarine Skuas, 3 Long-tailed Skuas and 18 Sabines Gulls were seen throughout the day. St Ives Island is not the only good viewing point, although the birds come closest here. Pendeen Watch is another good place, where the walls of the lighthouse provide some shelter but, alternatively, one can brave the gales and venture onto any of the exposed headlands along the north coast such as Gurnard's Head or Bosigran to find shelter among the rocks and enjoy the spectacle in blissful solitude. Seawatching from Gwennap Head in strong south or south-westerly winds can also be rewarding, but the birds are usually further out to sea. However, in July and August, many birdwatchers congregate here in the hope of seeing two species of large sheawater which appear in our waters during their non-breeding seasons, i.e. the Cory's Sheawater and the Great Shearwater. The former breeds on islands in the mid Atlantic and the latter even further afield, on south Atlantic islands. This peninsula is one of the best places in Britain for observing these birds, but they are usually only seen when a series of depressions causes a significant disturbance in the weather systems out in the Atlantic. Manx Shearwaters are fairly common, with thousands breeding on the Welsh Islands and small numbers breeding locally on the Isles of Scilly. The less common Sooty Shearwater is a summer visitor from southern oceanic islands.

An account of the seabirds would hardly be complete without a mention of the spectacular diving displays of Gannets, which take place when a shoal of fish comes close offshore. The incredible aerodynamics of this bird, as it folds its wings, turns and shoots like a torpedo into the sea with a deadly accurate aim, is both an admirable and an awesome sight. Compare this with the technique of the Storm Petrel which flutters above the surface of the sea, pattering its tiny webbed feet while it daintily picks off plankton from the surface, and you will begin to understand the diversity of bird life on the sea. A Gannet diving for fish is a common occurrence here, but a sighting of the little Storm Petrel requires time, patience and a good telescope. Neither of these birds are regular breeders here; the nearest Gannets are on Grassholm Island in Wales and there is a small population of Storm Petrels on the Isles of Scilly. In 1997 two old eggs of Storm Petrels were discovered on the Brisons, off St Just, and these birds are caught regularly on the cliffs on summer nights, although they are likely to be non-breeders.

Changing seabird populations usually means a decline in numbers, but there is certainly one success story in Cornwall and that is the relatively recent colonization of the Fulmar. This has gradually spread down the coast from northern Britain, reaching Cornwall in the 1930s. Before that, only vagrant birds had been recorded here. One was seen at Carn les Boel near Land's End in 1934, and this was followed by further sightings in the area. It is now normal to see Fulmars cruising effortlessly round the cliff tops, making full use of the updraughts. They are a fine sight as they glide past on their long, straight wings, casting a sideways glance with their soft, black eyes as if they are being curious. Like Herring Gulls, they have pale grey wings and are white below, but they are totally different because they are petrels and not related to them. The tubular nostrils, which are obvious through binoculars when the bird is close, are diagnostic of the petrel family. Fulmars are well adapted for gliding and are superb when seen negotiating very rough seas, but when it comes to landing on their ledges, as they must do to breed, they experience great difficulty and usually miss several times before they get a secure hold. Their legs are used mainly for paddling along on the sea as they pick up fish offal and plankton, and not for walking or standing upright as gulls do. Research has shown that they do not breed until they are six to eight years old so that, common as they seem, only a small proportion are likely to be breeding birds. They are widely dispersed around the peninsula, but Land's End and Pendeen Watch are particularly good places to see them.

Puffins, Razorbills and Guillemots are members of the auk family,

Razorbill

and are referred to as auks when they are seen at sea but cannot be identified as particular species. Our populations of all three species have declined in Cornwall for several reasons, one of which may be a perfectly natural change in sea temperature that has resulted in the redistribution of the small fish populations on which they feed. Others such as oil and chemical pollution, overfishing, disturbance and egg-collecting are clearly man's responsibility. The latter activity took place on a large scale in the past, because seabirds eggs are very palatable. In the late fifteenth century one Stephen Hoskyn of Penzance paid Thomas Beurying approximately 1,444 Puffins as a form of rent and Richard Carew (1602) writes of puffin chicks being eaten in Cornwall. Today puffins breed in very small numbers on the Isles of Scilly and in North Cornwall but not at all in the extreme south-west, although they occasionally land on the Brisons, off Cape Cornwall, when they can only be seen through a powerful telescope! They have been seen carrying fish but breeding has not been proved. Puffins do not breed until four or five years old and non-breeders will occasionally bring fish to land. In May and June they are recorded regularly flying offshore, but only in ones and twos. They are seen much less frequently than Razorbills and Guillemots because, apart from immediately around their breeding colonies, they normally stay well away from the coast, even in winter.

The latter two species of auk are seen in the waters around Cornwall in large numbers for most of the year, although not for a while in the late summer and early autumn, when they are rendered flightless during their moult and stay out to sea for safety. This is when they are most vulnerable to oil pollution. Few breed locally. There are small colonies of Razorbills and Guillemots on the Brisons and on the Armed Knight, near Land's End, as well as scattered pairs on the north coast. They disperse widely after breeding, and the birds we see off our coast in the off-season could be from as far afield as Scotland, Wales or Ireland. The best way to see Razorbills is to view them from Land's End with binoculars, when they are occupying their rocky crevices on the Armed Knight during April, May or June. Early morning is the best time and, in March and April, they may be joined by Guillemots which take part in their display activity on the water, when the birds are strung out in a line with their stubby black and white bodies facing the same direction. During these displays, they dive in synchrony. This activity has a social function whereby it stimulates the urge to breed. Auks fly along a straight path with fast wing beats that require a lot of energy, but they also use their wings for 'flying' underwater and are superbly adapted for diving and swimming in pursuit of small fish.

Cornwall, to the outsider, conjures up an image of a rugged county with small fishing harbours, rocky coves and a wide expanse of sea. The sounds and smells associated with the sea help to make up the special atmosphere that is so much loved by holidaymakers. An important part of this 'atmosphere' is the presence of the seagulls which are very much taken for granted. Their raucous cries are heard everywhere and their bold, aggressive gestures become familiar as they steal titbits and fishy morsels from innocent hands. They can be a great nuisance when they build their nests on the roofs of houses or make too much commotion, but their success is partly a result of careless rubbish disposal by humans on the land and in the sea. Perhaps they should be given the benefit of the doubt, considering pet cats and dogs have equally antisocial habits.

St Ives has 'enjoyed' a long association with the seagull, and this was quite apparent to Hudson in (1908): 'All this noise and fury and scurry of wings of innumerable white forms, mixed up with boats and busy shouting men, comes to be regarded by the people concerned as a necessary part of the whole business, and the bigger the bird crowd and the louder the uproar, the better they appear to like it. For the gulls are very dear to them.'

Gulls are distributed round the coast generally, but large numbers of them can be seen around the fish docks at Newlyn and in the harbour area at St Ives. There are several different species that are common locally and they are easily distinguishable. The most common is the Herring Gull, identified by its pale grey back and flesh-coloured legs. It is closely related to (indeed some taxonomists will argue that it is the same species as) the Lesser Black-backed Gull. The latter, which is not common here during the breeding season, differs in having a darker grey back and yellow legs. These two will interbreed and produce fertile offspring, so that the 'same species' argument has considerable sway. Scientifically, they are said to be at the two ends of a genetic cline. The piratical and extremely handsome Great Black-backed Gull has a ruling authority over these last two gulls. It is larger, with an almost black back and flesh coloured legs and has a deeper, hoarser cry. Not only will this gull steal fish from other species, but it will sometimes kill birds only a little smaller than itself. I have witnessed Kittiwakes and Puffins being killed, and have heard fishermen tell of fights between two individuals resulting in the death of the loser. These are facts which may be difficult to accept, but it is as nature intended and the predator's role is to maintain the health and survival efficiency of the prey population in its natural environment.

The Herring Gull and the Great Black-backed Gull breed regularly on our coast, the former mostly on inaccessible cliff ledges and the latter on isolated stacks and islets, where they are safe from mammalian predators, including humans. Herring Gulls began nesting on roofs, first in St Ives in 1952 and later in Newlyn and Penzance. Recently, various methods have been tried to deter them in St Ives, where there once was a superstition that gulls were in some degree supernatural human beings, perhaps drowned mariners and fishermen returned in bird form to their former homes. In those days, the St Ives people had special protective feelings towards them.

Kittiwakes are small, elegant gulls with jet black tips to their pale grey wings and more slender, yellow bills. The juvenile birds are especially attractive because they have a striking, black W-shaped pattern on the wings. They have a graceful, buoyant flight and it is a delight to watch them flying above the huge, foam-flecked waves during a gale. The largest breeding colony, of about 300-500 pairs in the 1990s, was at Land's End where excellent views could be had of the birds on their nests from the new suspension bridge. This colony has now disappeared for inexplicable reasons, a trait not uncommon with this species. A smaller group nests on the north coast near Morvah. Around the colonies, and sometimes when they are fishing offshore, one can hear the unmistakable cries of 'kitt-ee-wayke, kitt-ee-wayke'. Climbing activities have sometimes caused disturbances to these and other breeding birds on the cliffs.

In the winter Black-headed Gulls and, to a lesser extent, Common Gulls are seen around the peninsula, as well as the occasional rare Iceland Gull or Glaucous Gull from the Arctic north. Newlyn fishmarket is the best location to see these striking white gulls. Sightings of Mediterranean Gulls are not infrequent today, but this may reflect an increasing number of experienced birdwatchers residing in the area.

Terns do not breed here but there are small populations of Common Terns on the Isles of Scilly. It is one of the highlights of spring when the first of these graceful birds, with finely tapered wings and a conspicuous, black cap, is seen flying offshore during its migration northwards. Sandwich, Common and Arctic Terns often gather in sandy bays like Sennen and Pednevounder and fish along the shoreline, making shallow plunge dives into the sea.

Hudson is unkind in his account of the Shag: 'John Cocking is the local name of the Shag, the commonest species of Cormorant on this coast, a big, heavy, ungainly-looking creature, the ugliest fowl in Britain, half bird and half reptile in appearance. On the water, where he spends half his time greedily devouring fish and the other half sitting on the rocks digesting his food and airing his wings...' although later he admits that, when engaged in the latter activity, they 'have a noble decorative appearance, like carved bird-figures on the wet, black, jagged rocks amid the green and white, tumultuous sea'. If the bird is ugly at all, it is certainly not so when viewed from the cliff-top, as it dives down from the surface of a crystal clear sea with its wings folded into the body. In good light it can sometimes be seen for a while as it snakes along, agile and fast, in pursuit of small fish. One theory states that it is important for the Shag to dry its wings by spreading them out because the plumage is lacking in waterproof qualities, and this can cause chilling and subsequent death. Shags nest in small groups low down on rocky ledges and in caves. Again, the viewing hut at Land's End, which is equipped with telescopes, is a convenient place from which to watch them.

Cormorants are closely related and extremely similar to Shags, so that the novice would have difficulty distinguishing between the two. One feature of identification is that the former, slightly larger, bird shows a white breeding patch on its flanks in the spring and summer months. There is a breeding colony close by on Mullion Island, off the Lizard peninsula. Both Cormorants and Shags were heavily persecuted in the past because they were thought to deplete the local fish stocks.

Their populations have recovered well.

Nearly every cove on the peninsula is the home territory of a pair of Rock Pipits. They are small birds with streaked, olive-grey plumage and, although rather plain, they are worth looking out for, if only to watch their charming parachute displays in spring, when they rise up into the air singing and then fall again slowly with their wings fully extended and tails cocked. They build their nests in dark, damp crevices low down in the rockface and feed on invertebrates, which they search for among the rocks or on the grassy slopes. They are very similar to, but slightly larger and greyer than Meadow Pipits, and their legs are dark instead of flesh pink. The call note and song is also similar, the former a sharp 'phist, phist'. There is a little truth in the claim of a fellow ornithologist that a bird flushed from the cliff-top will fly down onto the rocks if it is a Rock Pipit and inland if it is a Meadow Pipit.

No wading birds nest on the coast of this peninsula today, and the only clarified report is of Ringed Plover nesting at Sennen in 1924. Our beaches are much too disturbed by people and dogs in summer. However, many waders pass through on their annual migration, others overwinter and some non-breeding birds stay throughout the summer months. The black and white Oystercatcher, for example, is seen all year round, feeding on the rocks at low tide and joining up with others to roost communally at high tide. They are

Oystercatcher
Summer plumage

strikingly handsome, with bright red legs that match the eye-ring and the bill. With this latter implement they have a knack of dislodging limpets by hammering at their bases. In summer, young birds often retain the white collar that is typical of birds in winter plumage. They would often remain unnoticed on the rocks below the cliffs were it not for the loud piping notes which they make quite frequently. There are no records of them breeding here, but it is likely that they would if not for the excessive disturbance.

Turnstones and Purple Sandpipers overwinter on the coast and are seen regularly around Sennen Harbour and the Cowloe, as well as beside the Jubilee Pool in Penzance. It is best to look for them at high tide when they are unable to feed and are roosting together on the rocks just above the high water mark. At low tide, when they are moving around picking up small molluscs and crustaceans, they are remarkably well camouflaged. Common Sandpipers stop off regularly

Purple Sandpiper

during their spring migration and may overwinter here, too. Very often all that is seen of this species is the back view of a small brown and white wader flying round the corner of some rocks, with a characteristic, jerky flight and a high pitched whistle. When standing it displays a curious bobbing motion.

There are certain birds which nest on the cliffs and inland as well, for example some birds of prey and members of the crow family. Small birds inhabit the coastal scrub and heath and are also found in wild parts of the interior. The Willow Warbler and the Chiffchaff are locally distributed where tall scrub or woodland reaches the coast, as it does in some of the south coast valleys, or where there are tall hedges enclosing fields, as there are between Mousehole and Lamorna. Song is the best way to distinguish these two warblers. Whitethroats are generally common and widely distributed wherever there is low-growing scrub, especially gorse. They are rarely overlooked because they have the habit of perching on the top of a prominent twig, while they puff out their white throat feathers and sing out their scratchy but pretty song. These three warblers are summer visitors and most of them migrate south in the autumn to Southern Europe or Africa, where the climate is warm and food abundant. Some of the Chiffchaffs, however, remain in Southern Britain, especially in the south-west where winters are reasonably mild. Whitethroats, which winter in the Sahel district of Africa, suffered a severe decline some years ago when they were affected by droughts in this region of Africa, but they have since recovered and are very common today.

Walk through any stretch of mature heath, coastal or otherwise, and the chances are that you will encounter a Stonechat. The bold habits of this bird, including the readiness to scold anyone encroaching on its territory, ensure that its presence is well felt. The plumage of the male bird is striking, with black head, white collar and ochreous breast, and he makes a charming picture as he perches on the topmost twig of a flowering gorse bush and makes a sound like two pieces of glass being rubbed together. In the past these notes gave rise to the local name of 'Chakker Eythyn', Eythyn meaning furze (gorse). The female has duller coloration but, like the male, she displays conspicuous white patches on the wings in flight. The two severe winters of 1985-'86 and 1986-'87 nearly wiped out the entire resident population, but it has since recovered well.

A large expanse of open heath always has a low density of breeding birds. This is partly because the degree of exposure gives little physical protection, and also because heathland, generally, supports a limited invertebrate fauna on which most birds feed their young. Skylarks and

Meadow Pipits, both ground nesters, are well distributed when the heath is large enough to contain territories of a size necessary to support them, as on the coastal heath at Porthgwarra. The joyful song of the Skylark makes a wonderful contribution to the pleasures of walking the coast in spring and it is a sad fact that this species is in imminent decline.

The Wheatear has already been noted as a migrant, and it also breeds on the coast in very small numbers, choosing open places with rabbit-cropped grass and a stone wall or an outcrop nearby where it can select a convenient hole or crevice for nesting. Disused rabbit burrows provide alternative nesting sites. Borlase referred to it as the 'hedge-chicker' because of its harsh call, but the modern name, Wheatear, means 'white rump' which is its most striking feature when in flight. It has a bold, upright stance and bobs up and down in an exaggerated manner.

Three members of the crow family (corvids) breed regularly on the cliffs, and it is no wonder that the Raven, which is the largest among them, is depicted in old Cornish legends as being a manifestation of King Arthur, so noble and domineering is the character of this bird. His huge, black silhouette patrols the clifftop domain, where the nest, a massive heap of branches and twigs, is built on a concealed ledge. The Raven is distinguished from other corvids by its large size, shaggy throat and wedge-shaped tail. This latter character is a good way of identifying the bird in flight, although the croaking call, a distinguished 'cronk cronk', is quite unmistakable. Hudson claimed he was shadowed by a raven: 'He would fly up and down, then alight on a rock, a hundred yards away or more and watch me, occasionally emitting his deep, hoarse, human-like croak.' A count undertaken by myself in 1993 revealed twelve pairs nesting round the coast here.

His cousins, the Carrion Crows and Jackdaws, also nest on the cliffs, the latter in considerable numbers. The twangy cry of the Jackdaw is as much an integral part of the Cornish coast as are the wheeling Fulmars and the dashing flight of the Peregrine Falcon. In late autumn large, mixed flocks of corvids are present here, because our populations are supplemented by birds from the continent. In 1993, 3,500 birds, mostly Jackdaws, were recorded in the Land's End district and they stayed for several days before moving on. They were a truly magnificent spectacle, wheeling around in the clear, autumn skies.

The celebrated Cornish Chough has long been extinct in Cornwall as a breeding bird and vagrants are extremely rare. They are possibly wanderers from Irish or Welsh colonies or, more likely, escaped birds from Paradise Park in Hayle. In the early 1900s Hudson and Folliot-Stokes comment on its absence in the Land's End district, but Blight records them breeding near Porthgwarra in 1861 and Rodd received eggs from the parish of Zennor in 1852, where birds remained until at least 1870. That it was once fairly common in the county is indicated by the fact that some Cornish families bore it on their coat of arms, and there are also many written records going back to the fifteenth century. The sad plight of the last Cornish Chough was recorded in 1973, when a lone bird which had lost its mate patrolled the cliffs for some time searching for another of its kind. Penhallurick in *Birds of Cornwall* gives a long account of the demise of this species. The reasons for the decline are rather obscure, but the cessation of grazing on the cliffs, the taking of eggs and young when the population was at its lowest ebb, disturbance and the more recent use of agricultural pesticides are all thought to have contributed. It is sad indeed that this beautiful, regal-looking bird, with its shining black plumage and crimson-coloured bill and legs, no longer graces our coastline and its ringing cry has ceased to echo around the cliff-faces. Attempts at re-introduction are currently being made in Cornwall. As I write a chough of unknown origin has been seen several times on the cliffs between Porthgwarra and Land's End.

The most common bird of prey seen along the cliffs is the Kestrel. The hunting technique of this small falcon is to hover above the grassy sward, its sharp eyes intently focused on the ground below in order to detect any slight movement that might betray the whereabouts of an unsuspecting vole or shrew. Most people are familiar with the Kestrel because it has adopted the habit of hunting along motorway embankments. Out here on the cliffs, this handsome chestnut and grey bird can be studied at leisure in more congenial surroundings. Cliffs, quarries and derelict buildings provide suitable nesting sites.

The Peregrine Falcon almost became extinct in Britain through persecution and the accumulation of chemical pesticides in the food chain. For some years now, the use of these chemicals has been disallowed and the bird has received legal protection, with the result that it has made a spectacular recovery, much to the chagrin of local pigeon fanciers. A number of pairs now breed around our coast and naturalists and enthusiasts may once again witness the incredible power and speed of this large falcon as it dashes across the sky or stoops to give chase to its chosen prey. Pigeons are favoured and easy prey, but many other birds are taken, including seabirds. Threats of egg stealing and general disturbance still linger, but generally the prospects are good for the Peregrine and, apart from a few pigeon fanciers who have only their own interests in mind, most people are delighted to witness the comeback of this falcon. The peninsula now supports several pairs.

In the year 2000 birdwatchers came from all over Britain to see a Gyr Falcon, a larger, paler version of the peregrine which replaces it in the Arctic regions. This beautiful white falcon was seen for several days in the area of Cape Cornwall and the Brisons, where it was seen to take auks from the cliffs.

Buzzards have been common in Cornwall for centuries. 'Bargus' is an old Cornish name for Buzzard and Carn Barges is a common place name in the county. It is not an unusual occurrence nowadays to see one of these large, conspicuous birds biding its time, sitting on a telegraph pole or on a hedge, unperturbed by passing cars. On fine days in spring they like to make full use of the rising thermals, and pairs with adjacent territories may join up so that numbers of them may be seen in the air together, soaring and mewing in a general social gathering. They are quite at home on the coast and will occasionally build their huge nests of sticks on the cliff-face, although they nest more commonly in trees inland. When they venture out over the sea they cause a great disturbance and clamour among the gulls, which mob them persistently as they try to drive them away. No doubt the fluctuating rabbit population contributes greatly to their diet. Population estimates by Roy Phillips give a figure somewhere between thirty and forty pairs for the peninsula as a whole.

Twenty years ago, in the early 1970s, a pair of Little Owls frequented the wild open country behind the cliffs at Porthgwarra. By day they were seen hopping along the granite hedges and were well known among birdwatchers. Since then there have been few records and the species seems to have declined in Cornwall as a whole. Strange though it may seem, they are not native to Britain, the reason being that their spread was halted by the disappearance of the land bridge between Britain and the continent after the ice age. They were deliberately introduced in the 1800s and spread rapidly, even though they were persecuted by gamekeepers. In reality they are very useful predators of mice, rats and other creatures that are regarded as agricultural pests.

'Caseg coit' is an old Cornish name for the Green Woodpecker and its literal translation is 'mare of the wood'. The laughing cry or 'yaffle' does indeed bear a resemblance to a neighing horse. Its presence on the cliffs may come as a surprise, but since the late nineteenth century it has extended its range across more open country from its original woodland habitat. A rough stretch of coastline with an adjacent wooded valley is ideal habitat for this woodpecker as it combines good feeding ground and ample nest sites. After thirty years of residence in Penzance, E. H. Rodd reported only two sightings before 1864. Since then it has become established in the southern part of the peninsula and is occasionally seen on the cliffs between Mousehole and Porthgwarra. As it probes into the turf for ants or the like, it looks not unattractive but rather strange in its bright green and scarlet plumage among the mellow tones of the cliff vegetation. It is often seen clinging to a granite rock in typical woodpecker fashion, and when flying it has a characteristic heavy, undulating flight.

The multi-coloured and, particularly, the pure white feral pigeons, which inhabit large caves set deep in the cliffs, have an ethereal look about them, as if they don't quite belong to this rugged coast. Occasionally, an individual bird resembles the rare, once native Rock Dove from which feral pigeons are descended, but which are now extinct in Cornwall. These pigeons are breeding in quite different surroundings when compared with those of their inner city compatriots.

Mammals

The Fox has for centuries found a safe retreat from the huntsman out on the cliffs. It ranges throughout the entire peninsula; over farmland, moorland, woodland and the coast, but here on the cliffs it is probably most secure. To quote Carew (1565), 'The Fox planteth his dwelling in the steep cliffs by the seaside where he possesseth holds so many in number, so dangerous for access and so full of windings, as in a manner, it falleth out a matter impossible to disseize him of that his ancient heritage'.

Early morning is the best time to see the Fox out on the cliffs, and a handsome sight he is as he wends his stealthy way among heather and gorse out on to some rocky promontory where he will likely hesitate and glance at the sea before disappearing into a hidden den. I have seen one many times in the same place, stretching out on a rock to bask in the sun. Although the animal is in full view, the place is totally inaccessible and he is safe from everything but the gun. The bark of a dog fox or the eerie scream of a vixen breaks the silence of many a still winter's night in the wilder parts of the peninsula, where the sounds of urban society are distant. The Fox, being a wary and cunning predator, is generally disliked by poultry farmers or by those who breed game for shooting, but he has a place in our countryside and is a useful predator of rabbits. The population seems once again healthy after a recent outbreak of mange.

The otter population declined dramatically in Cornwall in the 1960s, as it did in the rest of Britain, and there are several reasons. The widespread use of organochlorine pesticides was certainly partly responsible, but river improvement schemes, otter hunting, disturbance and hard winters have been cited as other factors involved. However, the Eurasian Otter is now making a come-back and there have been recent sightings in the area. Otter spraint is found on many of our rivers and there have been several road kills of this elusive mammal.

The life history of the otter is beautifully and sensitively written in Tregarthen's *The Life Story of an Otter*, written in the early 1900s and based on this peninsula. This author has a most unusual empathy for wildlife and conveys wonderfully well the magic and fascination of watching this elusive mammal in the wild.

Not surprisingly, Common Mink have often been mistaken for otters because they share roughly the same habitat and are superficially similar. The basic differences are that an otter is larger, with a flatter, broader head and a tapering tail. That mink competed with and played a large part in ousting the otter population is disputed by most authorities, but even so it is undoubtedly a pest in our wetlands, preying on the eggs and young of native birds, particularly wildfowl. They first arrived in Britain from North America in 1929, when they were confined and bred on farms for their valuable pelts. This inevitably led to their escape, and they were first recorded breeding in the wild in 1956 in Devon. They increased and spread rapidly and, in spite of large scale trapping, there is still a substantial population in west Cornwall.

Our rocky coastline, with its concealed beaches and inaccessible caves, is ideal territory for the Atlantic Grey Seal. This is the only species present in the south-west because Common Seals only inhabit the

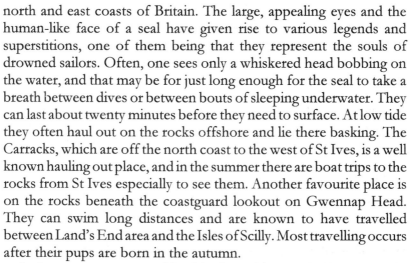

Atlantic Grey Seal

north and east coasts of Britain. The large, appealing eyes and the human-like face of a seal have given rise to various legends and superstitions, one of them being that they represent the souls of drowned sailors. Often, one sees only a whiskered head bobbing on the water, and that may be for just long enough for the seal to take a breath between dives or between bouts of sleeping underwater. They can last about twenty minutes before they need to surface. At low tide they often haul out on the rocks offshore and lie there basking. The Carracks, which are off the north coast to the west of St Ives, is a well known hauling out place, and in the summer there are boat trips to the rocks from St Ives especially to see them. Another favourite place is on the rocks beneath the coastguard lookout on Gwennap Head. They can swim long distances and are known to have travelled between Land's End area and the Isles of Scilly. Most travelling occurs after their pups are born in the autumn.

Seals can grow to a considerable size, with some bulls up to eight foot long. Their food consists mainly of fish, crabs, shrimps and mussels, but they have been known to take small seabirds. The males can be distinguished by their roman nose profile and are much darker than the paler-coloured females. Nowadays fishermen are more tolerant towards seals, but in the past they claimed that they threatened their livelihoods by competing for the fish and many were shot. Even today the seal is sometimes used as a scapegoat when the far worse problem of overfishing has led to the depletion of fish stocks. Having once been heavily persecuted, these animals are now legally protected and organised seal culls, like those on the Orkney Islands, rouse a great deal of controversy.

It would be a great omission not to mention the female Steller's Sea-lion that frequented the Brisons rocks, off Cape Cornwall, for many years. Binoculars or a telescope were required to see her and the best times were at low tide when she was often seen basking on the lower slopes of the rock. She disappeared in 1999 and presumably she is dead. The normal range of this largest of sea-lions is in the northern Pacific and how she came here is unknown, but it is possible that she was an escapee or was released.

Since Christmas 1991 children and adults alike have been able to watch the thrilling spectacle of a school of dolphins travelling and playing around our coast. Few people fail to be enchanted by their boisterous display of *joie de vivre* as they leap clean out of the water, sometimes twisting their gleaming black and white bodies upside down. They have been known to steal the show at the Minack Theatre and have been the highlight of a visit to the Land's End complex for many people. These lovable animals were remarkably bold and friendly at first, but they have since become more shy and retiring, possibly due to over enthusiastic boat owners chasing them. They comprise a group of about thirty Bottlenose Dolphins that split up into smaller groups for long periods, with the groups travelling farther afield than in previous years. Individual dolphins can be recognized by various scars or by fin shapes and one of them has had its dorsal fin bent right over since birth. This one is known as Benty.

Cornwall cannot compete with Scotland or south-west Ireland for the number of cetacean sightings but, even so, several species have been recorded off our coast, including Common, Striped and Risso's

Dolphin, Killer Whale, Pilot Whale, Minke Whale, Humpback Whale, and, more frequently, Common Porpoise. For the three years 1998-2000 small numbers (1-4) of Fin Whales have been present in our local waters and off the Isles of Scilly between the months of November-January, even entering Mounts Bay when fish shoals were evident. The twelve to fifteen foot spout of this whale is very visible. From January 2001, a Minke Whale began to be seen regularly in the Land's End area, a rather elusive animal, rarely seen in our local waters. It is well worth looking for cetaceans on trips to and from the Scilly Isles on the Scillonian, particularly in late summer. It is a sad fact that hundreds of porpoises and dolphins die by drowning in monofilament fishing nets and no solution has yet been reached to curtail this slaughter, which threatens the population of many species throughout the world. The Cornwall Wildlife Trust's group 'Seaquest' has an ongoing project, monitoring populations of sea mammals, turtles and basking sharks and is trying to identify and rectify any problems affecting the animals.

Basking Sharks, Sunfish and Turtles

In the months between April and October in 1998 and 2000 numbers of Basking Sharks (many hundreds) frequented the Cornish coast. They represent the largest numbers recorded for many years and the reason for their appearance on this scale is still unclear, although it is likely to be due to a prevalence of their planktonic food. Up to eighty-six were seen at one time in the bay between Carn Gloose, near St Just and Land's End, one evening in 1999. Obviously this was a most unusual and exciting spectacle for locals and visitors alike. The sharks may be up to thirty foot in length and there have even been reports of individuals comparable in length to a thirty-four foot boat. The animal is very impressive, with a large triangular dorsal fin, a thinner tail fin and a large gaping mouth that may be visible as it cruises along at the surface sifting food from the water. Some are recognizable as individuals by notches in the fins, fin shape or other body marks. A photo identification scheme is underway and from this, useful information on survival rates and movement can be gleaned. Anyone wishing to contribute records of sightings of these or other marine animals should contact the Cornwall Wildlife Trust at Allet. Tidal races offshore are particularly good places to look for the sharks. Sennen Bay, Pendeen and the Minack Theatre are recommended as popular haunts for the sharks. Breaching activity has been recorded frequently – the sharks may leap clean out of the water, a fact which surprises many visitors. Numbers peak in July and they are still seen occasionally into September and even October. Calm seas are best for locating them.

Sunfish are becoming regular visitors to our waters. They have huge, oval bodies but the only visible part is the thin, black dorsal fin which has a peculiar motion, flopping sideways down onto the sea's surface from the erect postion, always on the same side. At first sight, many people think they are seeing one flipper of a turtle! There has even been a report of a sunfish breaching. They are present in the summer months only.

Turtles are recorded in Cornish waters, most commonly the Leatherback, but also the Loggerhead and Kemp's Ridley Turtles. Breeding on tropical and sub-tropical beaches, they are not simply lost migrants, but they are animals which have undertaken a normal migration northwards to colder waters for feeding. They are regularly recorded and are part of our marine wildlife. There were several close sightings of Leatherback Turtle on the Scillonian pelagic trip in the year 2000. On that balmy summer's day in August one could well have imagined oneself in the tropics! Many of our records are from fishing boats and they occasionally become entangled in nets or ropes, but they may also be seen inshore. Leatherback Turtles may be almost two metres in length and may have come from breeding colonies in Central and South America or West Africa. The peak times for seeing them seems to be between August and October.

Passengers on the Scillonian 111 on the daily trips out to the islands always stand a good chance of viewing some of our marine wildlife closely.

*   *   *

Reptiles

The rough vegetation on the cliff slopes is good, safe ground for Common Adders, but apart from in spring when they are newly emerged from hibernation and the warmth of the sun is not great enough to quicken their response, they are difficult to find. One needs to be quiet and to move over the ground very slowly and stealthily, always looking ahead. Adders like to bask in the sun in sheltered hollows, and they often reappear in the same place day after day. Close up, they are revealed as very beautiful creatures, the males generally a greenish blue with strong, black, zig-zag markings, and the larger female, gold-brown with a dark brown pattern. The V-shaped marking on the head is a well known character of both sexes. Hudson was so charmed by the colour and pattern of one individual that he 'kept it for half an hour, carrying it to a piece of level green turf for the pleasure of watching the sinuous movements of so strange a serpent over the ground before I finally let it go into hiding among the bushes'. Such was the trust and confidence of this reputable naturalist.

There is little chance of misidentifying the adder because the only other snake that is common in Britain, the Grass Snake, has very

Common Adder

different markings and is much more scarce. Although Carew wrote of four foot monsters, the fact is that adders rarely exceed two foot in length. Now that fashions have changed it is safe to quote Folliot-Stokes who actually wrote of killing adders for their skins: 'Perhaps some of my readers would like to know that adder skins make an extremely handsome belt'. He follows this with details of how to procure and kill the adder and then how to make the belt.

Common Lizards also bask in the sun in order to increase their body temperatures sufficiently for them to be active. Again vigilance is required on the part of the enthusiast, to see more than a tail disappearing into the undergrowth. Their background colour varies from olive-brown to emerald green, but this last colour is quite a different green from the bright green of the rare Sand Lizard, which does not occur in Cornwall today although some old records may be authentic.

Fires on the moors and cliff slopes, accidental or otherwise, do a great deal of damage to all kinds of animal populations. Adders, lizards and slow-worms are particularly vulnerable and many are burnt alive before they can escape. The incidence of fire is slightly lower on the cliffs than it is inland so they are safer and therefore more common on the coast.

Butterflies

Many butterflies are attracted to the wild flowers on the cliffs and among them are some of the species that are common visitors to gardens. These are mostly the large, colourful butterflies like Red Admirals, Painted Ladies, Peacocks and Small Tortoiseshells. The first three are migratory and can sometimes be seen flying straight in off the sea as tiny specks of bright colour on the vast expanse of blue. That these fragile insects negotiate, at great risk, such immense distances in unpredictable weather conditions, shows how tenuous is their hold on life. However, enough must migrate successfully to continue the habit. Painted Ladies migrate from North Africa but they cannot survive the winter here, and it is the early migrants that breed to produce a native population in Britain in the summer months. The early influx may be spectacular with tens of thousands of individuals flying in off the sea. They are often associated with the migration of

Silver Y moths. Red Admirals and Peacocks do hibernate but their populations are supplemented by immigrants from the continent.

Clouded Yellows are also migrants and, being such a bright, sunny yellow, they are very conspicuous as they fly around the cliffs. Because they are migrants, the first ones of the year are usually seen on the coast. They are common in some years and scarce in others and spring arrivals will breed and produce a second brood. The pale form *helice* has been recorded here many times.

One does not have to be a butterfly enthusiast to be thrilled with a sighting of the large, flamboyant Monarch butterfly. This is a rare vagrant from North America or the Canary Islands that sometimes gets blown well off course during its annual migration, which covers thousands of miles. It cannot be overlooked as it is about twice the size of a Peacock and its large wing span (110mm) causes it to flap and glide like a bird. In autumn 1999 many people visited Cot Valley where a Monarch stayed for several days feeding on Buddleia bushes and that particular year many individuals were recorded here and on the Isles of Scilly.

It is difficult to distinguish one species of fritillary from another and, as it is seldom possible to see them at close range, it is more a question of knowing which species occur here and in what habitat. The resident, but now rather scarce, Dark Green Fritillary is a large, fast-flying butterfly that appears orange in flight, but is actually delicately chequered orange and black above, with a green wash on the undersides of the hindwings. The large, orange Oak Eggar moth can be mistaken for a fritillary, because it also flies fast and erratically during the day. The flight period of this fritillary is July and August, a little later than the first brood of the locally common Small Pearl-bordered Fritillary which emerges in June. The small size of this last species is enough to distinguish it from the Dark Green, but it is only the position and number of silver pearls on the hindwing which separate it from the extremely similar Pearl-bordered Fritillary. This latter species may be extinct on the peninsula today, although two were recorded at Carn Gloose in 1999 and it is very possible that it is overlooked. There has been a general decline in the populations of both of these butterflies in Britain partly due to shadier conditions in woods, farming intensification and land drainage. The abundance of violets on the coast no doubt helps to sustain the populations of the Small Pearl-bordered Fritillary, since they are the food plant of the caterpillars. One other fritillary, the Marsh Fritillary, is resident here, but it is extremely rare, with only two known colonies near the coast. It is possible that it is found inland but is overlooked.

Marsh Fritillary

Brown butterflies on the coast are represented by the Meadow Brown, Wall Brown and Grayling. These species are not very brightly coloured, but they all have noticeable eyespots on their upper forewings. These serve to divert the attention of a predator away from the truly vulnerable parts of the insect's body, so that a bird will stab at the eyespot first, allowing the butterfly time to get away. I have caught Meadow Browns that show clear beak marks around the eyespot.

These species are easily distinguishable using a field guide and, in the case of the Meadow Brown, it is easy to tell males from females by the larger size and the orange windows in the upper forewings of the latter. Graylings are found on short heath, and are fascinating because of their superb camouflage, aided by the intricate pattern on the closed hindwings and the fact that the body, on alighting, is tilted at such an angle that it does not make a shadow. They are more localised than the other two species, which are relatively widespread on the coast and inland.

The Small Heath is similarly marked with an eyespot, and is common on coastal heath and rough grassland, as well as inland. It always settles with its wings closed, exposing the grey-brown underside of the hindwing and part of the orange upperwing with its grey border. The food plants of the caterpillar are bents and fescues, both fine-leaved, common grasses.

High summer sees the emergence of the Large and Small Skippers, aptly named because of the skipping nature of their flight. The Large

Skipper is distinguished by the presence of some light patterning on its wings, particularly on the underside of the hindwings; this is absent in its smaller relative. The caterpillars feed on Yorkshire Fog, Cocksfoot and other coarse grasses so that they have a wide distribution in the peninsula.

The Common Blue butterfly lives in discrete colonies, where its food plant, Birds-foot-trefoil, grows among cropped grasses. There is no doubt that the lovely, pastel blue colour of this and other species of blue butterflies makes them among the prettiest to watch. Several 'blues' flying together over waves of deep pink heather is a magical sight. It is remarkable that the iridescent blue is not a pigment, but is actually produced by the microscopic corrugations on some of the scales breaking up the light falling on them, as would a thin film of oil.

3
MOORLAND

Treeless, strewn abundantly with granite boulders, rough with heath and furze, the summits crowned with great masses of rock resembling ancient, ruined castles . . .

W. H. Hudson (1908)

This description by Hudson conjures up a general picture of the moor as it was then and how much of it still is today, although EEC grants, available in the early 1980s, encouraged the reclamation of certain areas for agriculture, particularly in the east of the peninsula. A large part of the Penwith Moors now enjoys ESA status (Environmentally Sensitive Area), which gives it some protection, and fairly recently the National Trust has resumed ownership of parts of the moorland area. The Cornwall Wildlife Trust owns Bosvenning Common, which is just outside Newbridge.

The moors stretch from east to west, between St Ives and St Just, and consist of a tract of rough, open ground, with several high crags, such as Carn Calver, Hannibal's Carn and Watch Croft. Carn Galver was described by Blight as 'a bold and curious pile of granite rock, about 623 ft. above the level of the sea. With the golden furze, purple heath, whortleberry, and the bright mosses and lichens on the rocks, this cairn has in colour a gorgeous appearance.' As previously indicated, furze is the old name for gorse. These descriptions and others, together with old photographs, show that the dominant vegetation cover has altered over the years in that much of Carn Galver is now colonized by bracken. Lack of grazing, fires and the fact that bracken is no longer cut are the most probable reasons for this change. Bracken was once widely used for bedding, packing apples and as a base material for the building of a hay rick. Its spread on the moor is fairly general, but lack of soil depth, poor drainage and exposure are factors which limit its growth.

By definition, moorland is simply uncultivated hill land, but locally the term is also applied to wetlands like Chyenhal Moor or Kerris Moor, which are floristically very interesting, and are considered in the chapter on wetlands.

Moorland differs from heathland in that it occurs in areas of high rainfall and poor drainage, which leads to the formation of a layer of peat. Permanently wet and acid conditions inhibit the action of bacteria which break down plant remains, and this results in an accumulation of material over the years, i.e. a dark brown, fibrous layer known as peat.

In the past this peat or 'turf' figured largely in the lives of the local people because they were dependent on this and furze as their only source of fuel for warmth and cooking, until the railways brought coal to Cornwall in the early 1900s. It was not good peat here though, according to Hudson, 'In some parts of Cornwall they have good peat called "pudding turves" which make hot and comparatively lasting fire. In the Land's End district they have only the turf taken from the surface, which makes the poorest of fires, but it has to serve.' Bread was baked in a round baking pot that stood on a polished stone on the hearth with smouldering turves built up around it and heaped on a flat lid. In *A Cornish Farmer's Diary* by J. Stevens, written during 1847-1918, there are many references to turf paring and furze cutting. From this fascinating account of life in those far off days, we gather that turves were cut in July and then allowed to stand in stacks for drying. In places on the moor, traces of ridges are still visible where turf-paring was carried out, such as on Chykembro Common.

Furze was collected for animal feed or fuel and apparently seed was sown to encourage its spread. Photograph 98 in *Early Photographs of the West Cornwall Peninsula*, compiled by Reg Watkiss, shows a woman at St Just tidying her furze and turf ricks at the beginning of this century. Fire was prevented at all costs then. It is only comparatively recently that 'swaling' or 'firing' has been carried out by farmers to encourage the growth of 'sweet' grass for grazing animals. Fires are very harmful to the environment if they are too extensive, too frequent or if they occur at the wrong time of year, i.e. late spring or summer. Only a few cattle graze the moors today, and even Hudson, in 1908, commented that 'The donkey is the only domestic animal to be met with on the hills around Zennor, cows and sheep being occasionally seen'. The Worgan Survey, published in 1811, states that farmers in West Penwith used the moors as pastures for their cattle from November until their calving, after which they were rented out to the poor who paid their cow rent in milk and butter. Some of the moor in the ownership of the National Trust is grazed today because of agreements with local farmers and limited grazing is often beneficial for conservation purposes.

Of further historical interest are the old trackways and paths which traverse the moors. Some of the more substantial ones lead to and from the various mine-workings, the most famous being The Tinners Way which leads from St Ives to St Just, but many are associated with mines on the moor itself. There are also paths to the well-known archaeological sites, like Men-an-tol and the Nine Maidens Stone Circle, or to the tops of the higher carns. People are free to wander at will on the moors where the freedom and space of this historical landscape acts as a wonderful tonic.

Moorland is currently of little economic value and, as Britain has enough land already under agriculture to provide food for the population, the future of the moors is fairly safe, particularly as they are designated as an ESA. The land is of great value both to conservationists and archaeologists. It is impossible to walk anywhere on the moor without some sign of man's ancient past. 'Belerium is haunted by the vast ghostly multitude', were the words of Hudson, and such a place is ideal for contemplative thought on the lives of these men who lived alongside and in harmony with nature. The modern attitude of 'control and exploit' is inevitably unhealthy for ourselves as individuals and for mankind as a whole. Old settlements, stone circles and many other ancient monuments are scattered across the entire region; they are well-documented and make fascinating study.

* * *

Plants

The question often arises at to whether the moors were originally wooded, but the general opinion among the experts is that the higher parts of Devon and Cornwall were always devoid of trees, and in all likelihood this included much of the Penwith Moors. If there was any major woodland clearance, it probably took place before and during the Bronze Age and not, as is often thought, more recently. It is generally accepted that certain monuments, like stone circles and standing stones dating from the Neolithic to the Bronze Age, were placed so that they were within sight of each other, implying that there were no trees around to obscure the view.

Isolated blackthorn, elder and ash trees struggle to survive where a little shelter exists, perhaps in the lee of a stone wall or a rocky knoll. They are pruned by the wind until they assume unusual shapes, and these stunted specimens help to illustrate how unsuitable are the conditions for the growth of trees on the exposed moor, where they are constantly battered by the strong, salt-laden winds. Buds on the windward side of the tree are killed, while the ones facing the leeward side grow out in an exaggerated manner, giving the tree its characteristic shape of leaning away from the prevailing wind. The trees may be distorted and bent, but they are beautiful as well, because they are moulded by natural elements and are integrated into the natural lines of landscape. It is sometimes possible to raise broad-leaved trees by growing a shelter-belt of fast-growing pines, as demonstrated in some of the large gardens, or on a grand scale in Tresco gardens, on the Isles of Scilly.

Our moorland vegetation is typical of the type that grows on poor acid soil and is dominated by heather, gorse, moor-grass and bracken, with a variety of associated species within these communities. Over the years there have been changes in the dominance of one species over another, as already cited in the spread of bracken

European Gorse
Ulex europaeus

across ground that was once dominated by heather. Turf-paring, furze-cutting, swaling or local climatic changes have all influenced the vegetation, but the appearance of much of the moorland today is the result of general lack of use. On the whole there are very few landscapes in Britain which have not been shaped or altered by man.

Gorse has a patchy distribution; both species are present and may form an impenetrable mass as far as we humans are concerned, but together with other rank scrub, it provides safe cover for foxes, rabbits and many birds. Flowering Common and Bell Heather turn the dry slopes and ridges into glorious carpets of pink and purple in late summer, while damp flushes and boggy patches blush pale pink with Cross-leaved Heath. With the dying of the heather and bracken come the most beautiful tawny colours, heralding the onset of autumn.

Lousewort and milkwort add rich dots of colour to the moorland vegetation, while another common plant, Tormentil *Potentilla erecta*, forms a low cushion of pretty, lemon-yellow flowers set in a mass of dark green leaves.

After a fire the moor looks black and devastated, with the charred skeletons of gorse bushes standing above the ashy debris. Sometimes, however, light burning induces the germination of flowering annuals from seed which may have lain dormant for years, or which could have blown in, and this may be accompanied by the growth of other plants which are normally checked by the dominant species. Many of these are widespread opportunists, like Foxgloves and Red Campion, others are the normal heathland associates and include the Heath-spotted Orchid, which can throw up many spikes a year or two after the community is opened up like this.

Lousewort
Pedicularis sylvatica

Other plants which benefit from fires are the moorland grasses, hence the practice of swaling to encourage young growth for stock grazing. Swaling is not necessarily good for conserving heathland because, in some places, grass has become dominant at the expense

Red Campion
Silene dioica

of heather, due to too-frequent burning. Two species of grass are very common on the moor. One is Bristle Bent *Agrostis curtisii*, a fine-leaved, grey-green grass, which grows in tufts and turns a lovely pale gold that shimmers and ripples in the winter sunshine. The other is Purple Moor-grass *Molinia caerulea*, which is a large, tufted grass with broad leaf blades that are tinged purple like the flowers, giving it its name. It, too, is attractive in the winter, turning a mellow, straw colour. The beauty of the winter landscape is due not to these colours in isolation, but to a patchwork of blended colours, namely gold, russet, green and purple-grey.

Bracken is generally disliked because it is so rampant, but it has its finer points. In its early stages it is a delicate shade of spring green, and then later on, when the brittle, dead fronds are bathed in light from the setting sun, it is transformed into glowing sheets of bronze-pink. Hudson was struck by the effect of a winter rainstorm on the bracken: 'The rain-soaked dead bracken has now opened and spread out its shrivelled and curled-up fronds, and changed its colour from ashen grey and the pallid neutral tints of old dead grass to a beautiful deep rich mineral red, which transforms it from invisible lace rags to these red fabrics of curious design spread upon the monotonous dark green bushes like deepest red camellia or reddest serpentine on malachite'.

Many plants take advantage of the light available early in the year before the bracken grows up. Bluebells, Wood Anemones, Dog Violets and Celandines are among them. Ground Ivy *Glechoma hederacea*, which has purple flowers, and the cream-coloured spikes of Wood Sage *Teucrium scorodonia* are less well known.

Except in the extreme west of the peninsula, Bilberry *Vaccinium myrtyllus*, referred to as whortleberry in Blight's description of Carn Galver, is not uncommon, but its glaucous blue berries are seldom

found among the lime-green foliage.
The reason for this is obscure, but
since they are very palatable, they
may be taken by birds as quickly
as they ripen.
Old walls trail informally
across the neglected landscape,
adding great character and
charm. Their purpose is never
clear, but they probably once
marked boundaries or enclosed field
systems. Sometimes a snake-like tangle of
ivy stems grows over the stones, with hardly a
leaf to show. This detail was carefully observed
by Hudson: 'The stems which are not thick, are

Ground Ivy
Glechoma hederacea

smooth and of a pale grey colour, and grow in and out of the crevices, and cross and recross one another, fitting into all the inequalities of the stony surface, and in places where they cover the wall, looking like a numerous brood or tangle of grey "serpents".' No doubt this upsets the stability of the walls at first, but in its latter stages of growth, it probably holds them together!

These old walls become encrusted with a fuzzy growth of grey-green lichen, giving them a wizened look. These lichens thrive especially well in the clean, damp atmosphere of Cornwall, as do the attractive ferns which grow in cracks between the stones. Black Spleenwort *Asplenium adiantum-nigrum* and Polypody *Polypodium vulgare* are the two commonest ones, while the rarer Lanceolate Spleenwort *Asplenium obovatum ssp. lanceolatum* is fairly localised. Also on the walls are neat clusters of shiny, round leaves about the size of old pennies, which belong to the Wall Pennywort *Umbilicus rupestris*, and which produce tall, cream flower spikes.

The stony nature of the walls, with their little pockets of shallow soil, create a similar type of habitat to the rocky outcrops on the coast, and therefore can support the same lovely combinations of plants such as Wild Thyme, English Stonecrop and Sheepsbit. These same plants may also grow on the peculiar granite piles that outcrop on the moor. In a damp, shady crack on the summit of Carn Galver the rare

Wilson's Filmy-fern *Hymenophyllum wilsonii* has been recorded.

In early autumn, when the first gales have begun to turn the tips of the bracken brown, most flowering plants have gone to seed. This is

Devil's-bit Scabious
Succisa pratensis

Wall Pennywort
Umbilicus rupestris

when the blue, button-like flowers of Devil's-bit Scabious *Succisa pratensis* appear, and they are welcome splashes of colour among the jaded tints of a dying summer. It grows on the damp, open moorland, and is of special value as it is the food plant of the scarce Marsh Fritillary butterfly.

MOORLAND 63

A tall plant with very exotic-looking, pink flowers is found in association with streams on the moor, particularly in the area between Zennor and Morvah. In common with certain other alien plants, it looks attractive in gardens, but once it has escaped, it becomes an aggressive colonizer that does not mix sympathetically with our native flora, and therefore detracts from the special character of the local landscape. The Himalayan Balsam *Impatiens glandulifera* now forms

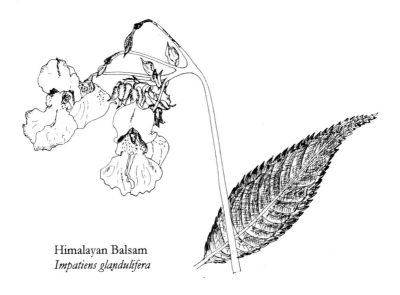

Himalayan Balsam
Impatiens glandulifera

large, uniform stands which are still spreading. Touch the seedhead when it is has dried out and there will be a mini-explosion of seeds flying in all directions; nature's perfect propagating machine.

Montbretia *Crocosmia x crocosmiiflora* is another alien which is becoming very common in the peninsula in many different habitats including moorland. Very often, its introduction is the result of the casual dumping of garden waste and it is a great pity that some gardeners do not feel the same kind of respect for the countryside as they do for their own gardens. The flame-coloured flowers are very attractive, coming as they do late in the year when there are few bright colours, but they have no place among the subtle colours that make up the landscape at this time of year. Its showy, light green leaves stand apart from our native shades of green. Montbretia, a hybrid from South Africa, now grows all over the peninsula, and large patches of it may be seen on the moor beside the Newmill to Gurnard's Head road, as well as in places on the coast. Attempts have even been made to 'beautify' the cliffs by planting it!

It has been fashionable for large estates all over Britain to grow Rhododendron *Rhododendron ponticum*, particularly along walks and bridleways. It thrived particularly well on our acid soil and has since begun to spread away from estate gardens onto the moor by seeding and suckering. It is abundant around Madron, especially at Trengwainton Carn and in the upper part of the Newmill Valley. Young plants spring up many miles from their original source, even way out on the open moor behind the Galvers and Watch Croft. A stand of Rhododendron supports very little wildlife and has no commercial value apart from firewood, and so it is not popular with conservationists or foresters.

The lack of diversity of plants on the moor is compensated for by the feeling of space in this open vista of everchanging colour. Like the vast, reflective sky above them, the moors alter greatly with the changing season and the weather and, for the sensitive observer, there are many different 'moods' to experience. Indeed, these moors, culminating in magnificent sea-cliffs, were to Folliot-Stokes 'the most arresting of all Cornish uplands'.

Birds

The moors are not rich in breeding species of birds, especially when compared with woodland or even gardens, because of general lack of cover and invertebrate food, as already explained. In summer the breeding birds are sparsely distributed, but in winter, although vast areas are devoid of any bird life, many find advantage in feeding and roosting communally, so that one may come across large flocks of them, in particular of Meadow Pipits or Linnets. For birds that stay all year round or for the summer or winter only, the expanse and

MOORLAND

Merlin

Raven

continuity of the moor are often very important factors in their survival. This is certainly the case with one of our winter visitors from the far north, the Hen Harrier, which patrols a large, well defined territory between November and March. The Penwith Moors only support a small number of these large birds, perhaps three or four, and they are often seen quartering walls and banks in their search for prey. They usually fly low, with a buoyant flight that is characteristic of harriers. The males are pale grey above and white beneath, with jet black wing tips and a conspicuous white rump that is diagnostic of the species. With such superb plumage, he is a magnificent sight floating over the heather, negotiating the banks and carns. The females and immatures are brown, but also show a white rump, and as they are virtually indistinguishable from the scarce Montagu's Harrier, they are collectively known as 'ring-tails'. While superficially similar to buzzards, harriers are slightly larger and have longer, straighter wings. Patient watching from a high vantage point, such as Carn Galver or Mulfra Hill, on a crisp, clear winter's afternoon, is the best tactic for seeing a harrier winging its way slowly and purposefully across the wintry landscape. They may follow the same route daily, which means they often appear at more or less the same time in a certain place each day. They sometimes roost communally on the ground at a chosen site.

Short-eared Owls sit tight in rough vegetation during the daytime, and will rise reluctantly when disturbed by a walker. These beautiful owls with golden patches on their wings are also winter visitors, but only in small numbers. On their breeding grounds further north the rate of success depends partly on the availability of prey, which includes small mammals like the lemming, so that in good breeding years there are more owls wintering here. Rodd (1880) states that they were to be 'generally found among heather and furze on the hill-sides, as well as in turnip fields', and this suggests that they were once more common than they are now.

Rabbits, voles and other small mammals and birds attract a variety of predators to the moor throughout the year. These include Kestrels, Buzzards, Ravens, Merlins and Peregrine Falcons. Rarer visitors such as Black and Red Kites, Hobbies and even a Booted Eagle have been recorded on the Peninsula, often lingering for days if sufficient prey is available.

After the autumn rain leaves water lying in the flats and hollows, Common Snipe move down from the colder north to winter on our moors in the mild south-west, where the mud is still soft enough to probe for invertebrates. These birds used to be known locally as 'heather bleaters', and the reason for the 'heather' in the name is most likely because they are frequently flushed from the grassy dips between the heather clumps. However, the call they make when they are flushed from the ground and fly off in their curious zig-zag fashion is more of a rasping note than a bleat. On their breeding grounds they have a call which they make from the ground that could be described as a bleat, or there is the familiar drumming produced by their tail feathers during flight displays which is likened by some people to a goat bleating distantly. This may indicate that they used to breed here, although the only record of this is from the Madron area before 1864 (Rodd). The fact that it was recorded at all suggests that they may have been a rare breeding species. When the bird is squatting on the ground, the brown, streaky plumage gives it an excellent camouflage. So long as the bird remains motionless it stays undetected, but when it feels threatened by the approach of an animal it will rise abruptly and noisily. If by chance the bird is silent on rising and has a shorter bill, then it is almost certainly the less common Jack Snipe. When an opportunity arises for close examination of the plumage, it is found to be exquisitely marked, like many wading birds'.

Woodcock arrive some time in late October to overwinter on rough, scrubby ground on the peninsula, including parts of the moor. They are closely related to snipe but are not so common, and indeed their numbers have been declining steadily. They too have cryptic plumage that is barred and marbled in fantastic shades of rufous, grey and yellow, which blend perfectly with the colours of the dead vegetation. Unfortunately, they are very palatable and it is a great pity that these beautiful birds are still shot. Early (and modern) literature tells us that when this highly esteemed meat is cooked 'the entrails are not drawn, but roasted within the bird, whence they drop out with the gravy, upon slices of toasted bread, and are relished as a delicious kind of sauce'.

When the rest of Britain is in the grip of a hard winter with several inches of snow on the ground or the earth frozen solid, mild temperatures often prevail in the south-west. This brings an influx of birds to the district from the north and east, a phenomenon that is known to birdwatchers as a 'cold weather movement'. The birds are naturally seeking food and warmth. They will stop here if conditions are better, or carry on south across the English Channel to France or west to Ireland, providing they have enough bodily resources. I have witnessed wave after wave of birds pouring in from the coast, driven on a bitter wind, many of them only to perish from lack of food. Lapwing, Golden Plover, Woodpigeons, Stock Doves, Starlings, Redwing, Fieldfare, Chaffinches, Meadow Pipits and several other species make up these flocks, which descend on the moor and the adjacent farmland, boosting the numbers of birds higher than at any other time of the year. Among these may be birds which winter regularly on the east coast of Britain such as Snow Buntings and Bramblings. Every winter will see some kind of influx, but total numbers will remain low when most of Britain experiences a mild winter.

At one time Lapwing and Curlew bred sporadically on the moor and there are records of both species from Lady Downs on the east side of the peninsula. There are no recent records, which is a great pity because their evocative calls and charming flight displays would do much to enhance the wild feel of the landscape.

Nearly every gorse thicket, stone wall, old building and rock pile is the home territory of the Wren, making it impossible to venture far on the moor before you hear its loud, scolding call, or see the bird creeping mouse-like in and out of the stones or vegetation. It is a resident as well as a winter visitor, and even when there are no other birds about, there is always a wren. It is a tiny bird with a big voice that lets you know quickly and sharply which parcel of moor is his.

Isolated patches of scrub, especially brambles, attract other common birds in the breeding season, such as Robins, Dunnocks and Blackbirds, while large clumps of willow may be occupied by Willow Warblers.

Spring has arrived when Meadow Pipits and Skylarks are up and singing. After a wild and grey winter, few sounds are more welcome than this sweet, heavenly chorus. Their aerial displays serve to advertise territory and attract mates. The Skylark rises more or less vertically and ascends to such a height that it is often out of sight of the human eye, whereas Meadow Pipits rise not so high, but at an

MOORLAND

Wren

angle, and then fall with wings splayed out like a tiny parachute. On a fine day, it is worth taking the time to walk out and sit against the prehistoric stones, just to tune in to the birdsong, a music that is older than ancient man.

Some Meadow Pipits and Skylarks sit out the winter gales and rain on the moor, others disperse into the adjacent fields and others migrate. In late autumn there is a visible migration southward of both species, and this probably involves some of our local birds. The Meadow Pipit is one of the commonest passage migrants to be seen on crossings to the Isles of Scilly. The two species look quite similar because they are both brown and streaky, and their restlessness, together with their habit of moving along the ground among the vegetation, makes them hard to see properly. However, Skylarks are larger with a more upright stance and they have a prominent crest in the breeding season. The songs are also quite different. These two species illustrate how birds that are similar can co-exist in the same habitat, because their choice of nesting site and their feeding strategies differ very slightly. One grey day in winter Hudson observed a solitary brown pipit 'being blown around like a leaf, emitting its sharp sorrowful little call', and describes how the pipit was to him 'the spirit of the place'.

Several other small birds either breed on the moor or visit it at certain times of year. For example, Linnets nest in isolated scrub. They were once popular cagebirds because of their pretty, rose-tinted plumage and their sweet, lively song that is given to mimicry. The male is brightly coloured with crimson patches on the head and chin, while the female is much less colourful. In August, adults and juveniles join together in large, twittering flocks and move about over open ground. As well as being migrants, Whitethroats and Stonechats are also fairly common as breeding birds in areas of low scrub. In spring and autumn, Wheatears stop to rest and feed in open rocky places and on trackways during their northward migration. Another migrant is the Whinchat, which looks like a pale Stonechat, but has a broad, creamy eye-stripe and white patches on the side of the tail. This species breeds in small numbers on Bodmin Moor, and the males in summer plumage are especially handsome.

The high, stony crags are particularly attractive to two other migrants, both of which have a general preference for bare, stony places. They are Black Redstarts and Ring Ouzels; neither are very common, but both occur regularly in small numbers in early spring and late autumn. Some Black Redstarts overwinter here.

When many thousands of migrants are on the move in late October and early November, these high crags provide good vantage points to hear and to see this overhead migration. Birds generally move on fine, clear nights, so that the early morning hours are the best time to watch out or listen for them. Many fly so high that their presence is only detected by their faint calls, but by tuning in to these the experienced birdwatcher will be able to identify the species. The high-pitched squeaking of Redwing, as they fly over on bright starry nights, is a familiar sound to them. Migration is surely one of the most fascinating phenomenon in the study of ornithology: the sight and sound of all these tiny creatures flying hundreds and thousands of feet overhead, with an inbuilt sense of direction and purpose. Even today, much of the mystery of bird migration with its complex navigational mechanisms remains unfathomed.

Mammals

Rabbit grazing has played a part in maintaining grassland and moorland habitat for many hundreds of years, benefiting some wild plants and inhibiting the growth of others. The Normans introduced the Rabbit to Britain in the twelfth century, most probably for food and skins. It was originally from the Western Mediterranean area. At first the animals were confined on islands or in artificial, purpose built warrens and, although they naturally lived on the surface of the ground, they soon developed the burrowing habit. It was not long before they escaped into the wild and spread rapidly. They have been important commercially right up to the late nineteenth century and were a welcome source of meat for poorer folk who often had to resort to poaching in order not to starve. Finally these self-reliant animals, with their enormous capacity for reproduction, came to be regarded as pests when they developed a liking for cultivated crops and, in 1953, the myxomatosis virus was introduced with the result that 99% of rabbits in Britain were killed off. The survivors developed a resistance and today we have a lower, but fluctuating, population which responds periodically to new, virulent forms of the virus. Unfortunately, a new disease called rabbit viral haemorrhagic disease is now spreading through the population, having been transmitted throughout Europe from China via infected domestic stock. It has reached this part of Cornwall and may have very significant effects on the vegetation and also on the population numbers of its predators like the Buzzard and the Fox.

Many people, especially children, love the rabbit because it is such an endearing creature with a soft, cuddly body. It is not surprising, then, that a lot of effort has been put into breeding it for domestic pets or for showing. All these beautifully marked and multi-coloured rabbits originated from the wild stock, originally a Norman introduction.

Late evening and early morning are the times when the rabbit is most active. Fields, moors and cliffs come alive with their scuffling and scampering, as local farmers know only too well. Those who suffer considerable loss of young cereal crops and brassicas try to control them by shooting or ferreting. Predatory buzzards and foxes help to keep down the population in years when disease is not rampant. Indeed, the high rabbit population is most likely to be responsible for the remarkable success of buzzards in this area. Other predators include the elusive, and not too common, stoat and weasel.

There are also records of Brown Hares on the peninsula as recently as 1981, but I know of no sightings since then. Badgers' setts are found on the moor, but they are more common in woodland and scrub.

Butterflies

The Grayling butterfly, described in the coastal section, is sparsely distributed in this habitat, while the Meadow Brown, Gatekeeper, Small Heath, and Common Blue are common and widespread.

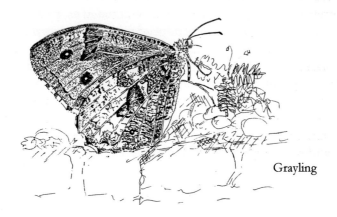

Grayling

Common Sorrel *Rumex acetosa* and Sheep's Sorrel *Rumex acetosella*, both associated with bracken, are the food plants of the Small Copper. This is a small butterfly which lives in discrete colonies on rough, open ground. It is delicately marked with black dots and borders on the bright copper forewings and copper bands on the dark hindwing. The metallic colour, as in the blue butterflies, is a product of scale structure and the resultant scatter of light. It has three broods during the year and can be seen as late as October. Reclamation of moorland and

MOORLAND

heathland in the past has led to a decline in numbers of this butterfly and certain others.

The scarce Silver-studded Blue occurs in small isolated colonies, particularly where there has been clearance or light burning, or where light reaches the ground between tall and leggy gorse. In these places gorse and heather bushes are in the process of regeneration and the young shoots provide food for the caterpillar, which will also feed on Bird's-foot-trefoil. The diagnostic, silver studs form part of the delicate pattern on the hindwing, more precisely in the outer row of black dots adjoining the orange band.

Between April and July the Green Hairstreak is on the wing, but it is easily overlooked on account of its small size and the fact that it resembles a brown moth in flight. This is because it is only the underwings that are green. It settles with its wings closed, exposing the lovely, iridescent green that is etched with a series of fine, white marks. The butterfly is found on heathy areas near the coast and inland, but it is relatively scarce. Its food plants include Common and Western Gorse, Dyer's Greenweed and Bilberry.

The Marsh Fritillary may occur inland on the moors where its food plant, the Devil's-bit Scabious, grows but at present the only known colonies are on the coast. It is an attractive butterfly, with a marked, chequered pattern, made up of darker and lighter colours than the other fritillaries. Fires are a great threat to this species, as it is not very mobile and does not easily recolonize.

4
WETLANDS

The typical Cornish bog, favoured haunt of the botanist, is not a thing of many acres open to the sky – it is a secret little place lying in a narrow valley, sometimes bordered by a copse and sometimes lying higher up the valley.
C.C. Vyvyan (1948)

The entire peninsula is well steeped in history and legends, and it is easy to imagine how they arose when one lingers in one of these 'secret little places'. Among the gnarled, lichen-encrusted willows, in an atmosphere heavy with the rich aroma of Water Mint *Mentha aquatica*, many a fanciful tale was born. It was said that the boggy area that makes up Woon Gumpus Common (previously known as The Gump), was a well-known haunt of the fairies and 'by night the miners crossed it in fear and trembling', wrote Blight in 1861. He also points out that in close proximity of The Gump is Kenidjack Carn, otherwise known as Hooting Carn, because of the 'ominous and fearful sounds the wind makes in passing round its jagged buttresses'. Tregarthen also writes of the 'evil repute' which clung to this particular area, and it was here that his earthstopper encountered a pure white badger.

There is nothing evil about the willow swamp surrounding Madron Well. This latter is supposed to possess miraculous virtues, and it is still visited by Christians and also pagans performing their ritualistic ceremonies. Nearby are the remains of a little baptistry.

Typical of Clara Vyvyan's bog is the secret, privately owned, willow swamp in the heart of Trevidder Moor, which is traversed by a series of half-collapsed wooden walkways, slippery with moss. In the water-logged interior, Water Rail and Moorhen creep between the soggy willow boughs and, in spring, the lovely, downward lilting song of the Willow Warbler issues from the leafy canopy, to join in a chorus with

WETLANDS

others of its kind. Trevidder Moor was once a favoured haunt of wintering wildfowl, but it has become more overgrown and wildfowl are scarce. These little pockets, of which there are many in the hidden valleys of the peninsula, are very rich in wildlife. Fortunately they are

Grey Willow. Male Catkins
Salix cinerea ssp. oleifolia

Grey Willow. Female Catkins
Salix cinerea ssp. oleifolia

relatively remote and undisturbed in this 'age of the motor car', compared with days gone by when, out of necessity, people crossed the countryside on foot on their way between farms, churches or villages. They must have known of the existence of these places, even though they would have skirted them. The importance of these sanctuaries within land that is intensively cultivated cannot be overemphasised, when loss of habitat is one of the reasons why so many animals and plants are threatened with extinction. Trungle Moor near Paul and Brewgate Moor at Sennen, both with a rich flora, were lost long ago. Chyenhal Moor, Clodgy Moor and Kerris Moor remain as good examples of wetland habitats, the former with a number of rare plants on record, but generally they are becoming very overgrown. Within the peninsula, many such areas remain only because they provide cover for game birds that are put down for shooting and also because they attract Common Snipe and Woodcock, both of which can still be legally hunted in the off-season.

Recently, many large ponds have been created by farmers for the dual purpose of irrigation and shooting. Domestic Mallard are released to be used as quarry and to act as decoys to wild duck passing over. These ponds can be of great interest, supporting many species of dragonflies and other aquatic life, as well as providing stopping off places for wading birds on migration in the summer.

In the past, local people were able to extract useful materials from these wetlands and therefore regarded them in a different light. For example, within living memory, willow was cut from Chyenhal Moor for making lobster pots. This prevented the moor from becoming overgrown and was, in itself, a good practice that maintained an open habitat at the same time as providing a useful and renewable natural material. On a small scale, reeds were cut and used for thatching. Slate roofs are now more convenient and less costly, while still being attractive and keeping the character of the old cottages. However, the traditional Cornish slate is being superseded by cheap, sombre black Spanish imports and reconstituted slate.

Moorland bogs are of a different nature, and they usually lie in the vicinity of a spring or follow the course of a stream. This description by Tregarthen could easily refer to the vast, boggy expanse below Boswarva Carn in the centre of the moors, north of Newbridge: 'Below them lay a stretch of marshy ground fed by some bubbling springs. Rills trickled along channels in the peaty ground, sparkling here and there between tussocks of rush and withered grass, losing themselves in a vivid green patch that fringed a chattering trout stream.' The Ordnance Survey map shows a fan-shaped network of small tributaries, rising to join the main river, which flows out to sea

Smooth contours of the granite cliffs above St Loy

The harsh outline of Mylor Slate at Carn Gloose

Old field systems at Zennor

Sea mist on the exposed cliffs at Pendeen showing the carpet of Thrift *Armeria maritima* and sea campion *Silene uniflora*

Wild flowers on a stone hedge on south coast cliffs

Wind-pruned hawthorn on exposed hedge

A Celtic cross with three-cornered Garlic *Allium triquetrum* growing around base, in a shelterd situation

Beautiful Demoiselle *Calopteryx virgo*

General distribution: English Stonecrop *Sedum anglicum*

Oxeye daisies *Leucanthemun vulgare*

Wetland: Cotton-grass *Eriophorum angustifolium*

Farmland: Corn Marigold *Chrysanthemum segetum*

Spring in Lamorna Woods with bluebells *Hyacinthoides non-scripta*, ferns and young Sycamore leaves.

WETLANDS

at Newlyn. Flanked by Boswarva Carn on one side and Boswens Common on the other, this place is a timeless wilderness, rich in plants and wildlife and, being part of an estate, is one hopes in safe hands.

Within the moor there are many small, accessible bogs in the valley-bottoms and basins, such as in Foage and Porthmeor valleys and that to the east of Men-an-tol where the main part of the Newlyn River rises. Water Shrews and Otters frequent Bostraze Common to the east of St Just, which is privately owned.

Plants

The tussocky nature of moorland bogs is due to water eroding channels through rush and moor-grass communities. The most common rush is Soft Rush *Juncus effusus*, recognised by its bright green, glossy stem that is only faintly striated, and the commonest grass is Purple Moor-grass. Gaining access into one of these bogs can be difficult because the irregular surface of the ground throws one off balance but, once in there, careful searching between the tussocks can lead to the discovery of some tiny, specialist plants. Even though many of these lesser-known, wetland plants are of diminutive size they are still a delight to behold. Blight was clearly entranced by the Pale Butterwort *Pinguicula lusitanica*: 'The flowers of a delicate pink colour are attached to slender stalks from two to three inches high; although this plant makes little show and may be passed unheeded, on inspection it will be found of the most elegant and graceful construction; the leaves are doubled up at the sides and veined in a curious manner. From their greasy nature the plant derives its name.' The pale green leaves, which form a basal rosette, are actually clothed with sticky, glandular hairs and, with these, they are able to trap insects. Although moorland bogs are its natural habitat, this plant has been known to colonise the sides of small drainage ditches dug through wet heath to improve the coast path by National Trust workers.

Sundew *Drosera rotundifolia* is another insectivorous plant, named either because drops of glistening dew become trapped in its leaves or, as the Rev. C.A. Johns claimed, 'the leaves have many pointed lobes or tentacles, each ending in a gland exuding a viscous liquid, especially when the sun is shining, so that they appear as if tipped with dew'. Insects are trapped within the leaves, which are covered in sticky, reddish hairs, and are almost devoid of chlorophyll. The tiny, white flowers hang limply from a very thin stem and rarely open properly.

Other plants found in these acid bogs include the attractive Marsh Violet *Viola palustris*, which has fine, dark veins on the pale mauve petals and Lesser Skullcap *Scutellaria minor* with a small, pink flower that produces an unusual looking fruit in that the sepals close over it like a lid. Bog Pimpernel *Anagallis tenella* has slender, trailing stems bearing pairs of tiny, rounded leaflets and funnel-shaped, rose-pink flowers standing erect on short stems.

Although many people take a delight in the large, showy exotics which grow so well here, those with an eye for a more subtle type of beauty will find more pleasure in searching for some of our rare natives, even if they are very small. Two tiny plants are not to be missed by really keen botanists. They are confined to damp, partially shaded places and are extremely localized, both with a very limited distribution in Britain. Wet banks or swampy open areas within willow carr (stands) are suitable habitats. They are the Ivy-leaved Bellflower *Wahlenbergia hederacea* and the Cornish Moneywort *Sibthorpea europaea*. Both were found in close proximity on Chyenhal Moor. Each plant has frail, entwining stems that sometimes grow over other vegetation for support or even over old tree stumps or stone. The bellflower has exquisite, bell-shaped flowers of a lovely, pale blue colour and neat, palmate leaves. I have seen it growing well on a damp lawn next to a stream, where it was mown over regularly. The pinhead-sized, pale pink flowers of the Cornish Moneywort are barely discernable, but on the tangled mass of thread-like stems are lots of rounded, delicately notched leaves that are about the size of an old sixpence and a vivid spring green in colour. Some of the granite stones of the old Madron baptistry were covered in it in the 1980s and, in

Marsh Violet
Viola palustris

1861, Blight recorded it 'trailing over the mossy stones by the rivulet of holy water', i.e. Madron Well, where it is most likely still found today. Cornish Moneywort is still locally common on the peninsula,

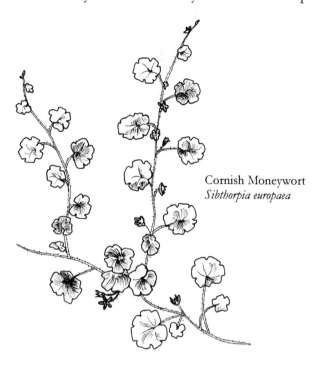

Cornish Moneywort
Sibthorpia europaea

but the Ivy-leaved Bellflower is becoming very scarce and, when reading past literature, one gets the impression that it has indeed greatly decreased. Perhaps, as well as habitat loss, it has been unable to tolerate raised pollution levels in the form of nitrates or other chemicals draining from the fields into valley-bottoms and into the streams by which it was once found.

While walking on the moor, it is easy to inadvertently stumble into a bog because there is often little indication of its existence, apart from the general abundance of Purple Moor-grass. However, there are certain plants with flowers that can easily be seen, giving due warning of a bog's presence, for example, the golden-coloured spikes of the Bog Asphodel *Narthecium ossifragum*, which is about nine inches high and consists of clusters of beautiful, star-shaped flowers and short, grass-like leaves that turn an intense tawny colour in the autumn. The unopened flower buds are tinged a delicate salmon pink. Another indicator plant is best seen as the early morning sun glances off its silver-white plumes. This is aptly named Cotton-Grass *Eriophorum angustifolium* and, according to the Rev. C.A. Johns, it was used for stuffing pillows under the name of 'Arctic wool'. Miniature forests of brilliant green Sphagnum Moss should certainly be avoided because such ground is treacherous and may even indicate a quaking bog.

In the middle of agricultural land lie Chyenhal, Clodgy and Kerris Moors. Reduced acidity, less exposed conditions and more variable soils, which may or may not be affected by run-off chemicals from the fields, result in a different selection of plants. Some places have become wholly or partially overgrown with willow, such as parts of Castallack and Chyenhal Moors, or they may be largely dominated by the Common Reed, like the upper part of the St Levan, near Porthgwarra.

Stands of Yellow Flag Iris *Iris pseudacorus* commonly occur in these wet flushes, particularly on the rich alluvial soil in valley bottoms. This handsome iris is a native, although the large, yellow flowers seem almost too exotic for our part of the world. The Rev. C.A. Johns claimed that the rhizomes yield a black dye and, if the seeds are roasted, they can be used as a substitute for coffee. In similar situations, fluffy, white, sweetly-scented Meadowsweet flowers *Filipendula ulmaria* mix with the stout, square stems of Water Figwort, while loose panicles of pretty, blue Brooklime *Veronica beccabunga* grow beside the lemon-yellow, buttercup-like flowers of Lesser Spearwort *Ranunculus ficaria*, named on account of its spear-shaped leaves. A less common relative of Brooklime, the Marsh Speedwell *Veronica scutellata*, is found at Chyenhal Moor.

Open, wet habitat may support Marsh St John's Wort *Hypericum elodes*, with its pale yellow flowers and grey-green, hairy leaves, or the straggling stems of Common Marsh Bedstraw *Galium palustre*, with flowers like tiny white crosses.

Part of the pleasure of botanising in these secluded bogs is appreciating the visual beauty of wild flowers, but many waterplants are strongly scented too, and, when the various scents mingle, the result is a rather special concoction. The smell of Water Mint is so strong and delicious that there is always a temptation to pluck a leaf and carry it about.

The Umbellifer family are often strongly scented and not always pleasantly. The Hemlock Water Dropwort *Oenanthe crocata* is one that some people may dislike. Pleasant or otherwise, the smell of this plant hangs heavy in the air wherever its robust stems have got a foothold. It stands five or six foot high in places and develops from sweet-tasting, but poisonous roots. Fools Water-cress *Apium nodiflorum* belongs to this same family and earned its name because it is often mistaken for

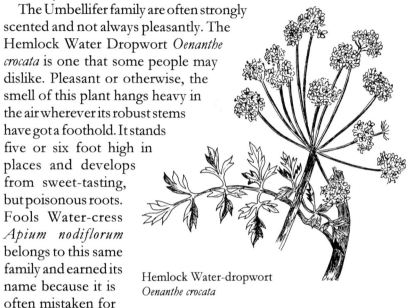

Hemlock Water-dropwort
Oenanthe crocata

Water-cress *Rorippa nasturtium-aquaticum*. This last species belongs to the crucifer family and differs from Fool's Water-cress by its solid stem, plain leaves with no serrations and flowers arranged in racemes rather than umbels. The two are often found growing together by streams, as they do at Penberth Cove.

The special charm of Ragged Robin *Lychnis flos-cuculi* is its unusual, ragged petals, the same deep pink as Red Campion, to which it is related. It is very localised and often found in wet meadows in association with Meadowsweet. Yellow Fleabane *Pulicaria dysenterica* and Cuckooflower or Ladies Smock, *Cardamine pratensis* are also common in these damp situations. All these plants occur together in several places such as Penberth, a marsh near Madron Well and Foage Valley.

Orchid lovers searching in the wetlands will soon come across the deep pink spikes of Southern Marsh-orchid *Dactylorhiza praetermissa*, which is distinguished from the similar Heath Spotted-orchid by the broad spur and generally unspotted leaves. The latter species is more

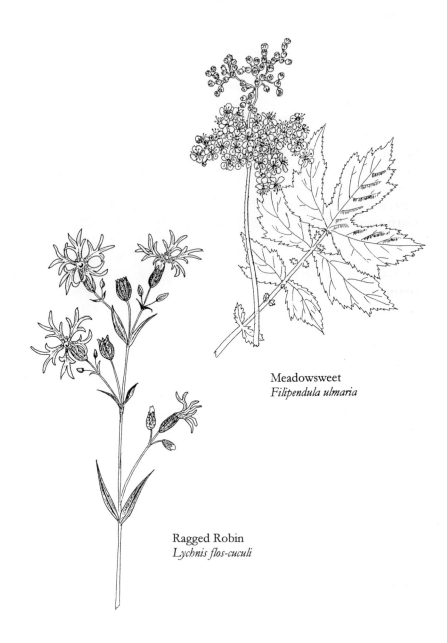

Meadowsweet
Filipendula ulmaria

Ragged Robin
Lychnis flos-cuculi

common and widespread. The Marsh-orchid flowers at the same time but normally grows in wetter situations, although I have seen hundreds of spikes come up in a fairly well drained field.

Permanent areas of shallow water often contain clusters of oval, long-stalked leaves which bear spikes of very small, green flowers in the summer. This is Bog Pondweed *Potamogeton sps*, a large group of plants in which it is hard to identify the particular species by leaf shape because this varies according to the depth and speed of water flow.

Another plant that is mostly immersed in water is the rare White Water-lily *Nymphaea alba*. This is a favourite introduction to garden ponds for its ornamental value; indeed, although it grows in a wild situation on the moor, it is unlikely to have a native origin here. A smaller, non-native, yellow Water-lily is found in a flooded quarry at Carfury, where another interesting introduction, the Water Hawthorn *Aponogeton distachyos*, produces unusual looking white flowers and resembles a large pondweed. Water-fern *Azolla filiculoides* once grew in this quarry, but it has since disappeared, possibly due to very cold temperatures. It still survives in a flooded pit near Ding-Dong, where it forms a mat of tiny, blue-green fronds that spread over the surface of the water and turn red in the autumn. This is an another alien and one which is very invasive.

The unusual growth form of Greater Tussock-sedge *Carex paniculata* is reminiscent of a Triffid in our science fiction films. Its large size makes it look even more formidable as it can grow up to four feet across and two feet high. These locally common sedges are only found in a few wetland areas now and the destruction of a large population during the construction of the Penzance by-pass gives the few remaining colonies in the peninsula a special value. Small populations occur on Nanquidno Downs and by a roadside bog, east of Sparnon.

A number of smaller sedges are found in wetlands, for example the very common Glaucous Sedge *Carex flacca* and Carnation Sedge *Carex panicea*. Then there are some difficult and obscure ones like Star Sedge *Carex echinata*, which is usually well concealed within a lush growth of moorland grasses and rushes. Sedges may be distinguished from rushes by their triangular stems. Of the latter group, three large ones are common; these are Soft Rush, Hard Rush *Juncus inflexus* and Compact Rush *Juncus conglomeratus*, while smaller ones include the

Toad Rush *Juncus bufonius* and Bulbous Rush *Juncus bulbosus*. For other species of sedges and rushes refer to the list of plants in the appendices.

Ferns have a general preference for damp and shady situations and are more or less restricted to woods or partially shaded hedges. A few grow by moorland streams and the Royal Fern, described in the coastal section, is one of these. In the valley bottom at Porthgwarra a large stand of this fern was burnt in the summer fire of 1995, but the new, copper-coloured fronds re-grew almost immediately, because the damp situation stopped the fire burning deeply into the soil. Hard-fern *Blechnum spicant* also grows on stream edges and ditches, although it is found predominately in woodlands. The rare Marsh Fern *Thelypteris palustris* was discovered on Clodgy Moor by the Cornwall Wildlife Trust, the only site for this fern in Cornwall, but it has not been relocated recently.

Drift Reservoir is one of the few sites for Bulrush or Reedmace *Typha latifolia*. The long chocolate-brown heads are familiar to most people from their popular appearance in flower arrangements, but the spread of the plant can be a nuisance. Its close relative, Branched Bur-reed *Sparganium erectum*, also grows right on the water's edge here.

Some wild plants can spread downstream by vegetative means, i.e. by the movement of fragments of roots or stems, while others disperse their seed in this way. Some garden plants can also do this, which helps to explain the spontaneous appearance of some wetland aliens a long way from the nearest gardens. One of these, Marsh Marigold *Caltha palustris*, is actually a British native, but is most probably an introduction here. It grows, for example, in streams at Penberth, Newlyn and St Loy. Another one is the exotic-looking Monkeyflower *Mimulus guttatus*, which has bright yellow flowers, blotched with red. This is sparsely distributed along streams, particularly at Kenidjack, Porthmoina

Bulrush
Typha latifolia

and Newmill. One can easily imagine the Giant-rhubarb *Gunnera tinctoria* in the rainforest of South America where it belongs, as it is a plant of the most gigantic proportions with huge leaves like elephants' ears that are covered in coarse hairs. It adds a touch of the tropics to many gardens that happen to be adjacent to streams and there are large stands of it at Porth Curno, Lamorna, Cot Valley, Zennor, and Penberth. Wrens have been quick to take advantage of the shelter it affords and have even been known to nest on its stems. When its leaves become dark brown and disfigured in the late autumn, or when it is frostbitten, it is indisputably ugly.

The ability of alien plants to establish themselves and spread has got out of hand where the Japanese Knotweed *Fallopia japonica* is concerned. It is the scourge of many English and Welsh counties today and has completely taken over some of the valley bottoms on the west coast, namely Cot and Kenidjack valleys. Efforts by the National Trust to eradicate it are underway and appear to be succeeding with continued cutting and the use of chemicals. It is also well established in the Newmill valley and in Newlyn Coombe. Very difficult to control without drastic measures, it has been, and still is, a serious threat to our native flora. The 'forests' of knotweed can grow to five or six feet tall and the plants bear broad, vivid green leaves and clusters of small creamy flowers which attract many flies.

Birds

Marshes, streams, reservoirs and ponds in the district support a few breeding species, but of greater interest to many birdwatchers is the number of regular migrants and vagrants which pause to rest and feed by water. These are mostly wading birds, although certain warblers, wagtails, herons, crakes and even some birds of prey are associated with waterside habitats. There are summer visitors that include nesting warblers, while some ducks and swans are here for the winter only.

The Grey Heron pays regular visits to many streams and pools in the district and is a familiar sight on the reservoir. Tregathen's friend, Ned, described the only sighting he had of this stately bird on the peninsula. It was apparently in a perfect setting, standing in isolation beside a pool on the open heath: 'A Heron in full breeding plumage standing still as a statue in the shallows of a sparkling pool. I remember how lovely he looked. It was on the moor above Lanyon Quoit, when the early furze was in bloom, and both the Heron and myself were after the trout.' There is only one record of herons breeding here and that is of one pair near Chysauster in 1970. Birds that visit the district are either non-breeders or birds that have wandered away from the vicinity of their breeding colony in the winter. Herons breed close by at Marazion.

For many years now, Little Egrets have been spreading northwards and westwards into Britain from the continent. The graceful stature of this small, heron-like bird, together with its snow white plumage, makes it delightful to look at and it is a welcome addition to our wetland fauna. In the early part of this century, large numbers of egrets were slaughtered for their feathers, particularly the head plumes. They were used for hat decoration, a diabolical trade set up to satisfy the whims of vain women. This eventually became a contentious issue and Hudson was one of the many inspired naturalists of the day who took it up. The controversy and the changing fashions finally put a stop to it. Nowadays, the egrets are present in ever increasing numbers in southern Britain, and they are now breeding in Cornwall. Sightings have occurred at various places on the peninsula, including Drift reservoir, and they have been seen on many occasions from the cliffs, flying along the coast. They are always present at the nearby Hayle Estuary and on the Helford and Fal rivers, especially in autumn when there is an influx of birds. Little Egrets are seen to best advantage when they perch in dead trees, when even the diagnostic, yellow feet can be seen! Trees growing close to water are used as nesting sites. In the spring of 2000 a rare Great White Egret from south-east Europe paid a short visit to Drift reservoir.

The same man who described the Grey Heron on the moor also had glimpses of 'what is most beautiful in our bird-life – say of a Kingfisher flying low over pools, when the sun catches its breast and feathers'. In the present time, this exotic-looking bird, which has the most brilliant turquoise and orange plumage, is only seen regularly in one place on the peninsula, i.e. Drift reservoir, where it still nests. Single, vagrant birds are seen from time to time in the summer around the coast as part of a post-breeding dispersal movement. It is odd to see a Kingfisher flying low and straight over the sea, as they are usually associated only with rivers. Although they are brightly coloured, they are small and sometimes it is only the high-pitched whistle it gives out in flight that draws attention to the bird.

Throughout Britain, pollution leading to decreasing fish populations, hard winters and the unsympathetic management of river-banks has brought about a severe decline in the overall numbers of Kingfishers. In the past, this incredibly beautiful bird was a popular subject for taxidermy but, thankfully, most people would now rather see them alive. An incident in the early 1900s, when the Mayor of Penzance ordered two to be shot when they were seen stealing fish from a pond in a town park, is one that is not likely to be repeated.

Large ponds surrounded by bushes, especially willows, are sure to be frequented by Moorhens. Sitting quietly in a well-concealed

Moorhen

position is the best way to observe them. Hudson gave an evocative description of the Moorhen in typical surroundings: 'Then came another sound, the sudden loud, sharp note of alarm or challenge of a moorhen yards away. There she stood on the edge of the water, in a green, flowery bed of water-mint and forget-me-not, with a thicket of tall grasses and comfrey behind her, the shapely, black head with its brilliant orange and scarlet ornaments visible above the herbage.' This shy, skulking bird is unlikely to be confused with the less common Coot because of the distinct, white bill and forehead of the latter.

Coot

Water Rails have such a skulking and secretive manner that they may be overlooked as breeding birds, but certainly most of the birds recorded here are migrants or winter visitors. Damp, well vegetated places by streams or ponds are suitable habitat for them. They are regularly seen around Drift reservoir and along streambeds, e.g. in Penberth Valley. They are capable of making some very strange, squeaking and grunting sounds which are mostly heard at dusk and in

WETLANDS 89

the early morning. Shaped like a slim Moorhen, they have grey and brown, streaky plumage and long reddish bills, and are unlikely to be mistaken for anything else, except perhaps their smaller relative, the Spotted Crake, which is a scarce migrant in Cornwall.

Two small birds are restricted to breeding in marshy places. The Reed Bunting is one of them, with the male the more striking of the pair in the breeding season, with bold, white moustachial stripes on its black head. Both sexes have streaked, brown backs and are paler beneath. They are unobtrusive, little birds with a simple, unmusical song that is easily overlooked. Some are resident but migrants and winter visitors also occur in small numbers. Sadly, they are a declining species. The other small, nesting bird is the Sedge Warbler which, regardless of appearance, does not go unnoticed because of its loud

Sedge Warbler

and distinctive song. This is made up of a repetitive garble of sweet and rasping notes, drawing attention to the bird more because of its unusual rather than its musical qualities. The only other local bird with a similar, but less harsh song is the Reed Warbler, which breeds occasionally at Skewjack, near Land's End. Both these warblers are small and brown, but easily separated by the very plain, unstreaked plumage of the latter.

Fast-flowing streams adjacent to open ground or a rocky beach are the favourite haunt of the Grey Wagtail. This most elegant of birds has lovely, contrasting plumage with a slate-grey back, pale underparts and bright yellow belly and flanks. Most noticeable is the incredibly long tail which it flicks as it moves from stone to stone. When disturbed, it gives a loud, metallic call that may be a single or double note. They most likely nest on the peninsula and are frequently seen by the streams at Penberth, Drift, Newmill and Lamorna. Preferred sites for nest building tend to be above running water, such as ledges beneath bridges, tangled tree roots or culverts.

Dippers, best known for their unusual habit of walking underwater to gather food, have recently begun to nest on our streams after an absence of many years. Many of the winter sightings could be local birds or they could be birds that have dispersed from other breeding sites in Cornwall, such as Bodmin Moor and the Camel or Foy rivers. In 1936 and 1937 a pair bred in Portheras Cove, east of Pendeen, but after this there was a long gap in the breeding records. They are delightful birds to watch as they stand bobbing on the rocks with white water gushing and spitting past them; indisputably handsome in black, white and rust-coloured plumage. The usual encounter with a Dipper is a glimpse of the rear end of the bird as it flies straight and fast out of view, with a parting whistle.

Drift Reservoir has attracted large numbers of wintering wildfowl since the dam was built in 1969 to flood the valley. The greatest numbers of duck are seen in January and February and consist mostly of the following species: Tufted Duck, Pochard, Teal, Wigeon, Shoveler and Mallard, while less common ones include Goosander, Goldeneye, Greater Scaup, Pintail, Gadwall, Gargany and Long-tailed Duck. These duck may also turn up in small numbers on some of the ponds dug by local farmers for shooting. Rare winter visitors

WETLANDS

in recent years have been a Ring-necked Duck and a Lesser Scaup, both from North America.

There are resident Mute Swans on the reservoir which have bred successfully recently. The real origin of this species is obscured by the fact that they have been domesticated in Britain since the twelfth century and most birds are now at least semi-domesticated. They are regarded as royal birds. Swans were kept on the Trengwainton Estate in the Victorian era until such time as the ponds became grown in. Whooper Swans come down from the north in cold winters and cause the strongly territorial Mute Swans to expend a great deal of energy in trying to drive them away. These wild swans from the far north have yellow and black bills, as have Bewick's Swans which are also scarce winter visitors. The latter are of slighter build and the detail on the bill pattern is different. Black Swan and Muscovy Duck, both escapes, have been recent visitors.

Other winter visitors include Cormorants, which fly inland to feed on fish stocks, grebes (mostly Little Grebes) and mixed flocks of gulls that come to bathe and preen on the water. Herons are regular visitors all year round, and naturally, the supply of fish is a great attraction for this bird and other species, too. The reservoir has a large population of Rudd, the origin of which is unknown, natural Brown Trout and stocks of Rainbow Trout. Fishing licenses are issued and it is a popular place for anglers. It is only a pity that a very small area of the reservoir is not set aside for a nature reserve, as this would allow certain species like Moorhen and Little Grebe to breed here. Disturbance is a crucial factor.

Birdwatchers scan the gull flocks hoping to find rare species and, quite often, Mediterranean Gulls are reported with groups of Black-headed Gulls. Bonaparte's Gulls, from North America, have been seen as well as Iceland and Glaucous Gulls.

In very dry summers, for example in 1995, the water levels drop dramatically, exposing large areas of mud. This attracts a great variety of migrating waders including common ones such as Greenshank, Redshank, Common Sandpiper, Green Sandpiper, Dunlin and other less common ones like Spotted Redshank, Ruff and Wood Sandpiper. Lesser Yellowlegs, Pectoral Sandpipers, and White-rumped Sandpipers, all American vagrants, have been recorded here in recent years.

One of Britain's most magnificent birds of prey, the Osprey, has occasionally stopped to rest and feed here during its migration to or from its winter quarters in Africa. Most years, birdwatchers have splendid views of one of these birds sitting in the treetops or catching fish. One hopes that the increase in breeding success of the Ospreys in northern Britain means they will be seen even more regularly in Cornwall in the future.

The reservoir has many different moods. In winter when a bracing, northerly wind whips up the steely blue water, it is difficult to pick out the duck bobbing on the white-streaked surface, but on calm, summer days when water levels are low, exposing tree trunks and walls, waders call while in flight, swallows and martins dip and myriads of insects haunt the edges. Man has created this habitat and nature has been quick to exploit it.

Dragonflies

A sunny summer's day in any kind of wetland habitat means that there are sure to be dragonflies and damselflies on the wing. They appear as soon as the sun has warmed their bodies sufficiently for activity; on cold, cloudy days they remain hidden in the vegetation. Most of those seen will be resident species, but rare migrants or vagrants do sometimes occur.

Few insects can match their incredible speed and manoeuvrability in flight as they dart, twist and hover in the air. They are fascinating to watch, as their flights are usually associated with specific behaviour, such as courtship, defending territory or the pursuit of prey. The combination of resplendent colours, delicate, translucent wings, streamlined bodies and huge compound eyes make this group of insects especially attractive. Their association with ponds and streams is vital, because they lay their eggs beneath the surface of water and the larval stages, which may take days, months or years to develop into adults, are totally aquatic. Although they are seen mostly near water, freshly emerged insects haunt dry places such as heaths, woods or along hedges for many days before seeking water to begin their breeding cycle.

Females of the Golden-ringed Dragonfly *Cordulegaster boltonii* are the longest bodied dragonflies in Britain, being 61-65mm in length. This species is very distinctive and most impressive, with a series of golden rings distributed along the black abdomen and thick, golden stripes on the thorax. This is a northern and western species in Britain, preferring fast flowing acidic streams and rivers. It is widely distributed in the peninsula and is often seen over dry land during its patrols.

The Emperor Dragonfly *Anax imperator* is also a powerful individual, but in this case it is more associated with ponds. It is an awesome sight as it dashes about faster than the eye can follow it and is a fearsome predator, capable of snatching a passing butterfly, biting off the wings in the air, and then devouring the body. Perfect, multi-coloured wings flutter down and lie on the water, a harsh lesson in nature. The predominant colours of this dragonfly are blue or green, depending on age and sex.

There are several species of dragonfly in which the males have chalky blue abdomens, while the females are generally brown, although old females may acquire the blue colour known as pruinescence. Details of colour, shape, wing venation, size, etc. are required for precise identification (see species list in the appendices). One species is only common when bare mud is exposed on the edge of a pond. This is the Broad-bodied Chaser *Libellula depressa*, a strongly territorial species that is fascinating to watch as it defends its section of pond or shields its mate from marauding males. Keeled Skimmers *Orthetrum coerulescens* and Black-tailed Skimmers *Orthetrum cancellatum* are similar in colour to this last one but different in proportion and other details.

Smaller and more slender than these two, the Common Darter *Sympetrum striolatum* is unmistakable because, apart from rare migrant darters, it is the only red dragonfly that is common on the peninsula, although in Britain as a whole there are several other species. Only the males are red; the females are brown. There is great excitement among local naturalists, including many birdwatchers who have extended their twitching activities to this group, when reports come in of rare vagrants. Usually the one in question is the Red-veined Darter *Sympetrum fonscolombii*, which is not unlike the Common Darter but is a brighter crimson red and, as its name implies, has wings with red veins instead of blackish brown ones. This last character causes the wings to shine faintly pink or blue. On a small scrape, excavated in 1999 at the north end of Drift reservoir, four Red-veined Darters appeared in the summer of 2000. In 1995, there was an influx into Britain of Yellow-winged Darters *Sympetrum flaveolum* from the continent. Some of these red dragonflies, with yellow colouring on the inner part of the wings, were seen elsewhere in Cornwall.

Like tiny, coloured darts, damselflies dash about over the water and in and out of the rushes, favouring these stout stems to pitch on. Even when settled, they are not easy to approach without them being put to flight again, so quick are they to detect the slightest movement. Sometimes they are seen resting on the water with wings held against the abdomen when it appears that they are dead, although they are certainly not. Three species of blue damselflies are common and widespread, i.e. the Common Blue Damselfly *Enallagma cyathigerum*, the Azure Damselfly *Coenagrion puella* and the Blue-tailed Damselfly *Ischnura elegans*. Adult males are not too difficult to distinguish, but females and immatures (teneral phase) are very tricky.

The male Emerald Damselfly *Lestes sponsa* is superficially like the Blue-tailed Damselfly, but actually belongs to a different family

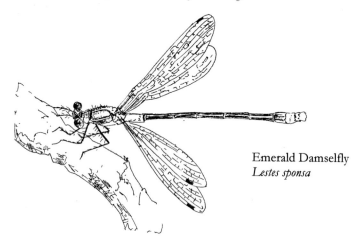

Emerald Damselfly
Lestes sponsa

altogether. It is a little larger and easily recognized by its habit of settling with its wings partly spread. The female is a beautiful metallic green. This species often pitches on willows growing on the edge of a pond. Large Red Damselflies *Pyrrhosoma nymphula* often fly at some distance from water. They are always the first to emerge, for the grisly reason that they prey on the newly emerged adults of later flying species. Both males and females are predominately red and, most importantly, they have black legs, distinguishing them from the rare Small Red Damselfly *Ceriagrion tenellum*, which has red legs.

The dainty, fluttering flight and the shimmering, metallic blue, green and brown colours of the Beautiful Demoiselle *Calopteryx virgo* make this damselfly the most eye-catching of all. It is found by streams where it patrols a chosen territory and pitches on tall vegetation growing along the edge. It is not uncommon on the streams at Lamorna, Kenidjack and Cot Valley. Indeed, it is well established in these valleys, where the streams are overhung by the knotweed, and frequently settles on the broad leaves.

In 1998, in late autumn, and at the same time as the arrival of several Monarch butterflies, an American dragonfly commonly known as a Green Darner made an appearance in Nanquidno Valley. Together with a few others of the same species, it had arrived in Britain as a vagrant never before recorded. Care has to be taken in identifying it as superficially it resembles an Emperor Dragonfly.

5

WOODLAND

Why do trees seem to be more nearly articulate than any other form of vegetation, and why is it that one can never tire of their company? Perhaps because their perennial attitude bears an age-old suggestion of aspiration and integrity. Yet who is more impressionable than a tree? To the very slightest modulation in wind or light weather, to the least mutation in the waxing or the waning season, a tree's leaves will respond by change in their movement or their colour.

C. C. Vyvyan (1952)

Many people have a special affection for trees and this windswept peninsula will be too barren and too open to please them immediately. However, there are many small pockets of woodland which are full of interest and have their own magic, although some of this may be due to the sharp contrast with the open landscapes of the moors and the rugged cliffs. One can imagine Clara Vyvyan's 'shadowy, green men' in the ferny glades of St Loy or grey-bearded gnomes hidden in the lichen-clad branches of the wind-pruned blackthorn at Porthmeor. The pure air and the high humidity in Cornwall ensure that a luxurious growth of lichens, mosses and ferns endow our woodlands with generous colour in the depth of winter and an exhilarating freshness when other foliage is sparse.

The Ordnance Survey maps of the district show a great deal of the woodland (or tall scrub) as ribbon-shaped strips following the river courses, particularly in the south of the peninsula, where it becomes more extensive in the coastal valleys of Lamorna and St Loy. Some of the valleys in the west and the north have groups of trees which can hardly be called woodland, but which are nevertheless interesting, as for example in Treveal and Foage valleys. Planting non-native trees

WOODLAND

for shelter and amenity on estate ground, as has been done at Trengwainton, Trevaylor and Boskenna, has given rise to blocks of woodland of a different character. At Hellangrove and Portheras, small plantations have been established as small commercial ventures. The origin of Kemyel Wood, on the cliffs between Mousehole and Lamorna, was discussed in the chapter on coastland.

The general lack of trees has been commented on already, but it remains to be said that after any Iron Age clearances there was a general depredation of the woodlands in Cornwall. In the sixteenth century Richard Carew wrote 'Timber hath in Cornwall, as in other places, taken a universal downfall, which the inhabitants begin now and shall hereafter rue more at leisure. Shipping, housing and vessels have bred this consumption, neither doth any man seek to repair so apparent and important a decay.' He also states, along with several other authors, that many trees were cut down for fuel and for use in the tin industry. How much, if at all, these activities affected our limited timber resources in the peninsula is not clear, but it is certainly likely that trees were taken down in the process of tin-streaming, when streams were diverted and a complicated system of banks and channels evolved. This kind of disturbance is very evident in some places, for example in the Lamorna valley and Skimmel Wood.

Plants

Scrub growing on damp ground in the high, uncultivated parts of the peninsula is most likely to be willow, while elder, blackthorn and gorse colonize the drier slopes. In the bottom of the south coast valleys, small blocks of mature woodland have developed. This once consisted of oak, elm and ash, perhaps with alder in the wetter parts, but many changes have taken place over the years. Loss of elm from the ravages of Dutch Elm disease, general felling and replanting has opened the way for the encroachment of other species, particularly Sycamore. This tree creates a lot of shade and builds up thick layers of slowly-rotting leaves that inhibit the regeneration of the native trees. This is best illustrated in the Lamorna valley where some large ash and oak trees still remain in the upper part of the woods but further down

Sycamore takes over. Horse chestnut, Sweet Chestnut, Holm Oak and Beech are among the trees which have been introduced, and these are seen particularly in St Loy woods. On some estates coniferous trees were planted, especially Monterey Pine and Monterey Cypress, when they were fashionable some fifty years ago. This is why we see many old, but occasionally splendid, specimens today, for example at Madron, Roskennals and above Newlyn. Both these trees are extremely salt-tolerant and it is ironic that the Monterey Pine, which reaches grand proportions here, grows nothing like so big where it is native on the Californian coast. Britain's only native conifer, Scots Pine, which grew throughout this country tens and thousands of years ago, does not do very well here but there are a few specimens which have attained a good size. Various types of firs, pines, spruces and larches are grown in the small plantations at Hellangrove, and also at Trewidden, Trengwainton and in other large gardens.

Of special interest locally are one or two specimens of the very primitive Maidenhair Tree, or Ginkgo, for example in gardens at Trevaylor, Trengwainton, St Ives and Alverton in Penzance. In some ways this tree is allied to conifers, but is deciduous and has broad, fan-shaped leaves that are reminiscent of the Maidenhair-fern. It is native to China.

The Evergreen or Holm Oak, often referred to as ilex, was introduced to Britain in the sixteenth century and is sometimes planted as an ornamental tree, mostly for its evergreen character but also because of its salt-resistant and wind-firm properties. It is a very handsome tree with finely sutured, grey bark and grey-green leaves that are reminiscent of the colours prevalent in its native Mediterranean region. Our evergreen Holly is a native and has a restricted distribution on the peninsula with a tendency to be more common on the eastern side. It sometimes makes a small tree. The waxy leaves restrict water loss, because the roots cannot take up water if the ground is frozen, and the prickles prevent animals from browsing when the ground is covered

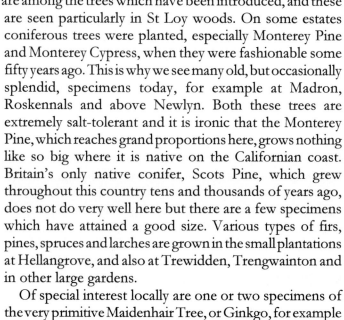

Buds: Sycamore
Acer pseudoplatanus

WOODLAND

by snow, although this would rarely apply in the south-west of Britain.

Other evergreens have been introduced and planted as windbreaks; among them are Tamarisk, Escallonia, Pittosporum and Veronica species. Small-leaved Pittosporum was once grown commercially and fractions of these small plantations still exist, for example at Boskenna and Lamorna. Certain varieties of Eleagnus are also planted for their wind resistance; some of them are very attractive with golden undersides to silvery green leaves and very fragrant flowers.

Evergreens are usually much appreciated because their leaves give colour in the winter months, but there is also great beauty in the patterns etched by the bare branches and twigs of deciduous trees against a bright sky. This caught the sensitive eye of Clara Vyvyan, prompting her to write: 'The trees cast off their coronals and stand up stark and unashamed; their boughs make patterns on the sky, and unencumbered, respond to every breeze, while underfoot the leaves have ended their susurrant melody and are now blended with the soil.'

There are other species of introduced trees that are not widespread but give character to certain places, such as the beeches in Trevaylor Wood, poplars at Bojewan, lime in Newlyn Coombe and hornbeam at Trevidder. Near Newlyn, at Stable Hobba, mature beech trees grow out of a small hedgebank, reminiscent of the hedges on Exmoor.

Buds : Beech
Fagus sylvatica

Many elms were affected by Dutch Elm disease; most of these have been felled, but a few have been left standing as stark silhouettes. Elms are a complicated family. Hybridization occurs between species and, indeed, the commonest elm in the county, which is often wrongly identified as English Elm, is considered to be a cross between

Cornish Elm and Wych Elm. Some expert botanists now maintain that even the Cornish Elm is not a native, leaving Wych Elm as our only indigenous species. Luckily, this elm is more resistant to the disease and there are trees of considerable size in St Loy woods. Leaf shape and general outline are the main criteria for identification.

The two locally common, native oaks, with their gnarled, mossy trunks and wayward branches, suit the landscape of West Cornwall where natural contours are broken and rugged. There are no smooth, rolling hills or broad expanses of flat land as in eastern England. Most of the peninsula is too windswept for tall, well-shaped oaks, but there are some large, mature specimens in the coastal valleys. Small, stunted oaks with twisted limbs grow out from hedgerows on the moorland edge, where they struggle to reach any size. Oaks will grow on the lee side of the moor as they do on Trencrom Hill where there are scattered, small, trees.

The Sessile Oak is more common on the lighter soils of western Britain, and it is this species that is most likely to be encountered here. The two main diagnostic characters, i.e. the stalkless acorns and stalked leaves, are reversed in the English Oak. Turkey Oak, originating from southern Europe, has been planted for example at St Loy, and one of the features of this tree is that the large leaves, which are bright orange when they have turned, stay on the young trees until the new ones form in the spring.

Hawthorn, along with European Gorse, Dog Rose *Rosa canina* and Blackthorn, grows in hedges and on the woodland edges. It acquires a lovely shape when it has room to spread and has been subjected to a little wind pruning. The flowers appear in May after the Blackthorn and, indeed, it is often called May. The blossom is faintly flushed pink and appears just as the bright green leaves are unfolding. The fruit of the thorn trees and the wild rose, sloes, hips and haws, are an important source of food for birds in winter. In poor berry years, the birds may go hungry, but less so in the south-west where the winters are not severe and alternative food is easier to come by.

Hazel, which is a fairly common, native undershrub in these parts, does not here produce nuts of sufficient size or quality for human consumption, but small mammals, especially grey squirrels, relish them. The wood was once a useful resource, providing thatching

Hips: Fruit of the Dog Rose
Rosa canina

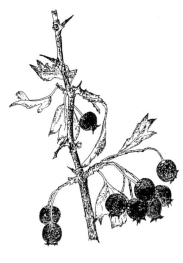

Haws: Hawthorn Fruit
Crataegus monogyna

spars, bean poles, hedge-stakes and firewood, while the catkins appeared regularly on the school nature table, and perhaps still do.

One of the first shrubs to come into leaf in the spring is Common Elder, the flower and fruit of which are popular ingredients for use in the home-made wine industry. Less well known is the art of flavouring pancakes by gently shaking fresh blossom over the half-cooked batter. The result is a subtle hint of the sweetness of spring. Elder is very common in the district, so it is not surprising that the old Cornish name for it, 'scawen', crops up in place-names such as Boscawen and Nanscawen. It is also noteworthy that badger setts are often associated with elder bushes.

Many of the planted woods contain garden shrubs like rhododendron or hydrangea, as do Trengwainton, Boskenna, Roskennals and Trevaylor. *Fuschia magellanica* does well in a salty atmosphere and so it is widely planted here. The exotic looking, pink and purple flowers appear in the autumn in many odd places. For example, fuchsia bushes are strewn among the blackthorn scrub in Treveal Valley and they also occur on the edge of Kemyel Wood. A small hedge of fuchsia grows against a low granite wall in the Penberth valley and the flowers

Blackthorn
Prunus spinosa

WOODLAND

look beautiful against the stone. Many unusual shrubs and trees have been planted in the woods here, some of which are tender and survive only because of the shelter and general lack of hard frosts. An isolated bay tree in St Loy, mahonia at Lamorna and a tree-fern in Skimmel Wood are a few examples. The latter may have propogated itself from nearby Trewidden gardens.

Up to the early twentieth century many apple orchards thrived in the area around Penzance, Newlyn and Lamorna and a local cider industry flourished here, hence the old presses exhibited in the town museums.

Hazel
Corylus avellana

Borlase wrote in *The Natural History of Cornwall*: 'Apples, pears and cherries were much cultivated and cider made'. Until a few years ago, there was an ancient apple tree at the entrance to Alverton car park in Penzance that served as a reminder of this era. Competition from elsewhere in Britain, along with an influx of imported apples, finally curtailed the industry. Some of the old Cornish varieties of apple trees have been preserved. Hidden in the woods at Stable Hobba is a small medlar, the fruits of which are said to be palatable only when half rotten.

Common Elder
Sambucus nigra

Late in January the first Snowdrop *Galanthus nivalis* appears in the woods, and although early varieties of daffodils from the Isles of Scilly have been in the shops since November, these fragile, white blooms are the first real sign of the yearly cycle beginning anew. Winter, albeit short, is on the way out. As protection against severe weather conditions, nature has provided snowdrops with hardened leaf-tips for pushing through frozen ground and flexible, elastic stems to bear the nodding flowers, enabling them to withstand strong winds. The inside of the corolla tube is delicately marked with a green ring and striations; a system which guides insects to the pollen clusters inside. A close encounter offers an opportunity to test the validity of Walter de-la-Mare's description of the smell: 'a faint, fresh, earthy scent, suggestive indeed of snow itself'. They were often planted around priories and other holy places because their whiteness was thought to symbolise purity. This historical fact, together with ornamental planting, makes the question of whether or not they have a native origin a difficult one to answer. Small beds of snowdrops are found in and around woodland in several of the valley bottoms, especially in the south of the peninsula.

Daffodil cultivars are not uncommon in some of the woods. Many flowers grow among the trees or within the remains of small, stone enclosures, as at St Loy and Penberth. It is said that old bulbs were sometimes thrown into the woods by farmers when they had no further use for them and they then became established, but on old estate land they may have been planted for ornamental purposes. Few daffodils have flowered in the last few years in St Loy woods, but there are always masses of blue-green leaves that look rather out of place among the natural green and brown colours on the woodland floor. The flowers in Kemyel Wood were shaded out long ago.

After the daffodils come the bluebells. Early morning is the best time to visit the woods, when the blue flowers are dew-laden and sparkling, and when the light glances off celandines scattered like stars on a dark background. Our moist woodland in the south-west is very suitable for bluebells and they are widespread so long as cattle are fenced out. At the top end of Lamorna woods, silky pink flowers of Pink Purslane *Claytonia sibirica* grow among the bluebells and the result is a charming mix of colour. This garden escape is found in only a few localities, but it is becoming more common.

The Three-cornered Garlic, already described in the coastal section, has spread into some of our woodlands and comes into flower before

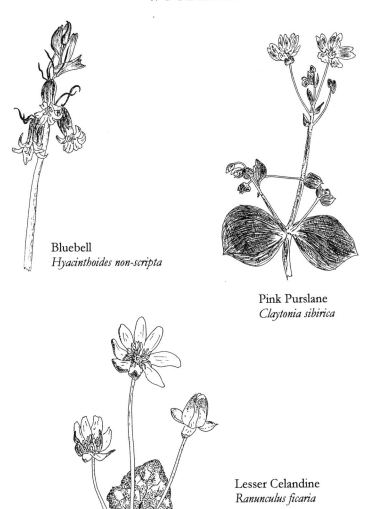

Bluebell
Hyacinthoides non-scripta

Pink Purslane
Claytonia sibirica

Lesser Celandine
Ranunculus ficaria

the bluebell. It does not seem to be so shade tolerant as the latter, but is still invasive to some extent. It is now well established in many other habitats and appears on the cliffs, on roadsides and in hedges. Coming originally from the Mediterranean to be introduced into gardens, it has naturalized itself only in the far west of Britain and Ireland. Both this plant and the Bluebell have tepals, which is the term applied when the sepals and petals are alike, and in this case the white tepals are striped green. Also in the lily family is the native Wood Garlic or Ramsons *Allium ursinum*, which has broad, spathe-shaped leaved and rounded clusters of white flowers. This species is only found in the east of the peninsula, especially at Newlyn Coombe. Again, it smells strongly of garlic. Two other members of the lily family, the Star-of-Bethlehem *Ornithogalum lusitanicum* and Solomon's Seal *Polygonatum multiflorum*, have naturalized themselves in St Loy woods and at Boskenna respectively, but they are very scarce.

Many other native plants make up the total flora of our woodlands; among them are Cow Parsley *Anthriscus sylvestris*, Dog's Mercury *Mercurialis perennis*, Enchanter's Nightshade *Circaea lutetiana*, Wood Speedwell *Veronica montana*, Ivy-leaved Speedwell *Veronica hederifolia*, Red Campion, Wood Avens *Geum urbanum* and Wood Sorrel. Yellow Pimpernel *Lysimachia nemorum*, Wood Sanicle *Sanicula eoropaea* and Early-purple Orchid *Orchis mascula* are less common. The extensive carpet of Tutsan *Hypericum androsaemum*, growing beneath mature willows at Roskennals, is possibly of native origin.

A plant which never fails to attract attention because of its curious structure is the Wild Arum *Arum maculatum*, also known as Cuckoo Pint or Lords-and-Ladies. The large, arrow-shaped leaves are fairly ordinary, but the tiny, purplish flowers are borne on a tall, spike-like structure called a spadix, which is itself partly encircled by a leaf-like spathe. The flowers are unisexual and consist only of the reproductive

Ramsons *Allium ursinum*

parts, i.e. stamens or stigma. This particular species is rare in the far west and is replaced here by the introduced Italian Lords-and-Ladies *Arum italicum* ssp. *italicum*, which is easily recognized by the cream veins etched on the shiny, deep green leaves and the yellow flower spike. Both arums are attractive plants, more because of their unusual structure. In autumn, the central stem bears clusters of shiny, red berries which are poisonous.

Dog's Mercury *Mercurialis perennis*

In the southern woods when the spring flowers are past their best, the trees come into leaf and the canopy closes. As the amount of sunlight on the woodland floor diminishes, the newly formed fronds of ferns emerge, golden tinted and curiously curled to protect the delicate tips. They slowly unfold and mature to become the dominant plants on the woodland floor. Their bright green fronds splay out from a central rootstock in elegant fashion, and once again the woods are green and luxurious.

Without too much difficulty and with a fairly comprehensive plant book, the different species of ferns that grow locally can be separated from one another. Fortunately for the botanist there is a restricted number of them. There are four common ones which have the conventional growth form described above. They are the Male-fern *Dryopteris filix-mas*, Scaly Male-fern *Dryopteris affinis*, Broad-Buckler-fern *Dryopteris dilitata* and Lady-fern *Athyrium filix-femina*. They are broadly similar and all have divided fronds.

The appearance of the simple, tongue-like fronds of the Hart's-tongue *Phyllitis scolopendrium* gives the fern its name and they are unmistakable, having a smooth, shiny, surface that is a lighter green than the other ferns'. It is a common fern of woodlands and shady hedges. Similarly, the dark blue-green, fronds of Hard-fern *Blechnum spicant* are easily identified because they are divided into narrow strap-like segments. This fern has the unusual character of bearing the reproductive organs 'sori' on special fronds which grow up from the centre of the plant and are narrower with finer divisions. In woodland, the Polypody often grows on the decaying boughs of trees where it looks especially attractive on the old wood among thick moss and encrusting lichen. The fronds of this very common fern are simply divided and are a bright, summer green.

It is only the semi-natural woodland that has a rich and diverse flora, with some of the original woodland plants still present. New plantations, especially coniferous ones, have a poor or even non-existent ground flora. The balanced ecology of natural woodland cannot be recreated once it has been destroyed. Many people do not realize the complexity and interdependence of an ecosystem that has taken thousands of years to evolve. Planting and replanting trees does not replace what is lost when ancient woodland has been destroyed.

Birds

Woodlands generally support large insect populations and so they tend to be rich in bird-life when compared with moorland or farmland. For many species of bird, breeding is timed so that nestlings are raised when certain insect prey is available in peak numbers.

Birds in woods are difficult to see in the closed canopy and it is useful to know the songs of the different species. Indeed, the best way to appreciate them is to visit the woods on an early morning in spring when the full chorus of individuals may be heard. They are fully occupied declaring territory and attracting mates. Much energy is expended in singing during this vital part of the breeding cycle and the woods are vibrant with a natural and beautiful chorus. Later on, when the eggs and young need protection, the birds cease drawing attention to themselves by singing and expend their energy on foraging for food to sustain sitting mates or young chicks as well as themselves.

There is a considerable variety of small birds breeding in our mature woodlands. The tiniest bird in Britain, the Goldcrest, is one of them. Hudson recalled some local boys calling it the 'Golden Christian Wrannie', the gold referring to the bright yellow crown stripe. It is a very pretty bird, mostly olive green with thin black and white striping around the eye and two small, whitish wing-bars. It is very good at hiding itself in the foliage, but its presence is easily detected once the high-pitched call or the tinkling song is learnt. It is not uncommon in our woods, particularly where there are scattered firs or pines, but it is nearly always difficult to see in the canopy.

Again, a preliminary knowledge of song is important when trying to distinguish between the warblers that co-exist in the woods. Visually, Willow Warblers and Chiffchaffs are extremely difficult to tell apart in the field, but their songs are totally different. The former species has a melodic, downward lilting song, while that of the latter is a repetition of two notes sounding not unlike 'chiff-chaff, chiff-chaff', hence the name. These two warblers are found in most woodlands and Willow Warblers will also nest in well grown scrub. The song of the returning Willow Warbler in spring is sweet delight to the ear of a keen naturalist.

Like Chiffchaffs, Blackcaps show a preference for more mature trees. They are unmistakable because the clear, black cap is quite obvious in the greyish coloured male and the female has a distinct, chestnut brown cap. The song is a rich melody, only to be confused with that of one other bird, the Garden Warbler. This latter species is a plain, olive grey bird which, although it is heard singing in the spring, is most likely to be a migrant passing through.

Tits, of which there are several nesting species, have a confusing medley of calls which will confuse and frustrate the most patient observer for a long time. There are five different species that breed and, in winter, they join forces and make up large foraging parties together with Goldcrests, an occasional Firecrest and overwintering Chiffchaffs and Blackcaps. Sometimes a wood, which appears to be totally deserted, is suddenly alive with the twitterings and light movements of birds making their way through the trees. Great Tits, Blue Tits, Coal Tits, Marsh Tits and Long-tailed Tits join up to form these mixed flocks. Coal Tits, like Goldcrests, have a particular liking for coniferous trees.

The less common Spotted Flycatcher is a scarce, summer visitor and a passage migrant. It likes to select a perch in an open glade or on a woodland edge, from which it can fly out into an open space to snatch insects from the air. Pied Flycatchers are migrants only and, with the former species, are seen regularly in trees and scrub within the peninsula during their migration.

Great Tit

Robins, Wrens, Blackbirds, Dunnocks and Song Thrushes contribute to the early morning chorus, all of them present in fairly good numbers, especially along the woodland/field interface. Currently, ornithologists are greatly concerned for the future of the Song Thrush, which has shown a severe decline in recent years. Groups of large, coniferous trees on the edge of farmland, for example at the top of the Lamorna valley or at Roskennals, attract Mistle Thrushes or Holm Thrushes as they used to be called. 'Holm' is Cornish for holly, and holly berries are eaten by these birds in winter. Mistletoe berries are eaten in Britain but not to a great extent, whereas abroad these thrushes are very partial to the berries of another species of Mistletoe that parasitises olive trees in the Mediterranean region, and this is maybe how they first became associated with the plant. The song is remarkably loud and rich and is often delivered from a desultory tree top, giving it its other name of 'stormcock'.

Loud resonant drumming, or a distinctive 'tchick', indicates the presence of a Great-spotted Woodpecker. When the trees are bare, there is a good chance of seeing this bird since it is made fairly conspicuous by its black and white patterned plumage, set off by a brilliant crimson crown in the male. Since the late 1800s, when its status in Cornwall was restricted to only occasional sightings in the east of the county (Rodd 1864), its range has increased enormously and today it is a regular nesting bird in many of our woods in the far west, for example at Skimmel Bridge, Penberth and Boskenna. The Green Woodpecker depends on mature, deciduous trees to find a

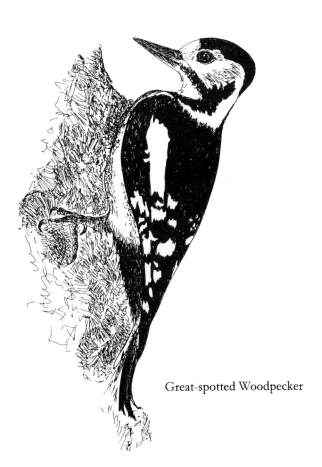

Great-spotted Woodpecker

Green Woodpecker

suitable hole in which to nest so that it frequents the woodlands as well as open ground. The Lesser-spotted Woodpecker is absent from West Cornwall but there are records of birds seen regularly for several years at Trevaylor Woods, near Gulval, between 1876 and 1883.

The Nuthatch, with its recently acquired habit of visiting bird tables, has spread further west since the days when Rodd (1864) described it as being almost unknown as far west as Penzance. This bird makes hammering noises up in the canopy as it endeavours to split open a nut jammed into a crevice. It is a colourful bird, with blue-grey upperparts and orange-buff on the breast and belly and is usually seen creeping round treetrunks searching for insects in the crevices. Like the Great-spotted Woodpecker, it is found in the proximity of large and mature trees. So too is the Treecreeper, a tiny bird with heavy streaking on its brown back and clean white underparts. It normally creeps like a mouse along branches or on the trunk of a tree, using its short, curved bill for probing into the bark for food items. Wherever there are large trees, even in the town parks, it may occur but it is not very common. It is seen, however, in some of the sheltered valleys of the south coast.

Woodcock fly in to roost at dusk on winter evenings and then, as night falls, the woods come alive with anonymous scufflings as various nocturnal creatures emerge. Tawny Owls hoot and flit across moonlit openings in search of small mammals and suddenly the woods are transformed into places of mystery and beauty.

The sudden screeching of an owl can be quite unnerving and, although our small woods still support Tawny Owls, they are seldom seen. It is only because they are so vocal that people are aware of their presence. The Barn Owl is different in that it is crepuscular, that is, it likes to hunt in the half-light of dawn and dusk. It is occasionally seen in woodland, if it is not too dense, and will nest in holes in trees. It was probably without sufficient evidence that Rodd (1880) claimed that Long-eared Owls were especially common in the neighbourhood of Penzance, e.g. at Trengwainton and Trevaylor, and in 1948 Ryves stated that there were only three known pairs in the whole of Cornwall. It is easy to overlook this bird, but at the moment there are no records of it breeding within the peninsula and it is unlikely that it was ever common. It is reported only as a scarce migrant here.

Buzzards frequently use trees in which to build their nests and there are records of a pair at Treveal that alternated between a cliff ledge and a tree site. The trees may be in woodland or they may be isolated specimens. Buzzards have always been fairly common in Cornwall, except during the nineteenth century when persecution by gamekeepers was at its peak, keeping the inland population at a low level, and also around 1955 when myxomatosis was rife among the rabbit population. Speaking out against persecution, Rodd (1880) claimed that when he examined a dead bird, he 'found in the stomach of one of these "inveterate poachers" three nearly full-grown rats instead of the expected remains of snipe, woodcock, partridge, hare or pheasant'. During the 1914-18 war, a reduction in the number of gamekeepers generally in Britain resulted in an increase of Buzzards and a return to areas where they had not bred before. Nowadays more people are interested in their protection than their slaughter.

A rookery in the peak of the nesting season is a place of great noise and commotion; birds fighting each other, stealing from neighbours and generally causing a tremendous uproar. For a time they disturb the peace and tranquillity of the neighbourhood but, in spite of this, they are generally liked and protected by country people, many of whom enjoy watching their curious and mischievous antics. The birds start nesting early in March when they break away from the large flocks of crows, rooks and jackdaws that roost and feed together in the winter months. They return to breed in their native rookery, in a small wood or in a group of mature trees. There are small rookeries at Sancreed, Lamorna, St Buryan and St Loy, among other places. Research shows that they have a preference for elms but, with the loss of many of these trees, they now use ash, beech and sycamore, apparently successfully. The rookery at Lamorna is well situated for observing the birds, being beside the local pub.

Unlike their gregarious relatives, Carrion Crows nest singly in a variety of habitat types, including woodland. As notorious thieves of eggs and young birds, they were once persecuted by gamekeepers, but numbers increased again when the two wars curbed the activities of these men who often did not bother to discriminate between rooks and crows, since rooks were believed to cause crop damage. Jackdaws are more numerous on the cliffs, but they also nest in holes in trees or old buildings rather than building an open nest. A fact which is often overlooked and for which they are seldom given any credit, is that all three species consume large numbers of defoliating insects, which is beneficial to woodland as well as adjacent farmland and therefore important in ecological terms.

Wood Pigeons were once restricted to woods but now they nest in hedges, isolated trees and, in other parts of the country, on the ground. Instead of relying on acorns and beech mast which is their natural food, they are now in the category of agricultural pests. In the southwest where farmers grow a lot of crops which remain green all winter, such as turnips, brassicas and clover, they have ample resources to exploit and are no longer checked by natural shortages. However, they are beautiful birds with dusky grey plumage, delicately tinged iridescent green and pink. Their soft cooing and flippant aerial displays are an integral part of the coming of spring in the southern counties. If they are disturbed while resting in the trees, they make a loud crashing noise as they fly through the branches in their haste to get away.

Mammals

The peninsula, with its many pockets of untamed land, be it rough moorland slopes, coastal strips or wooded valley-bottoms, supports a large Badger population. Until recently, compared with other parts of Britain, these animals have suffered little persecution save some control gassing in parts of Cornwall. However, as part of the MAFF experiment, researching bovine tuberculosis in cattle, many animals were trapped and killed in the summer of 2000. It is a controversial issue which at the time of writing remains unresolved.

Tregarthen gives a charming description in *Life Story of a Badger* based in the Land's End area in 1925: 'It is an ancient creature in an ancient land where the grey of its coat blends like that of no other animal so harmoniously with the grey of the rock, that only a practised eye can distinguish the wearer, when at dawn he threads the boulder strewn slopes or steals over the massed crags of a carn.' There are many established setts in woodland as well as on open slopes and even out on the cliffs. In the 1920s a count gave twenty-six setts between

St Ives and Newlyn on the coast and eighty-one inland, and MAFF officials have undertaken more recent surveys. Tregarthen, who was well-informed about the habits of wild creatures, gives their main food items as insects, roots, fruit, wasps, grubs and fish! He also notes that, locally, hibernation only occurs in severe weather. Unlike the earthstopper featured in his book (a character who was almost nocturnal himself), few people walk in the woods at night and so the animals themselves are seldom seen apart from on roads, and this includes the unfortunate ones which end up as casualties. In dry summers they may raid gardens because the earth becomes too hard for them to dig for worms, etc., and recently a badger gorged itself on a whole crop of strawberries grown by a friend and destined for eating during televised Wimbledon week! Badger watching involves positioning oneself upwind of a sett, preferably on a clear, moonlit evening. Patience is required; it can be a long wait until the first shuffling movements signify the animal approaching one of the exit holes. First it will sniff the air to check that the ground is clear and then, when it begins to emerge, the white stripes on the face become visible as it pulls its bulky frame from the hole. They use well defined pathways through the undergrowth and, not far from the sett, will be the badger's latrine. February and March are the months to look out for cubs.

Grey squirrels occur here in small numbers, mostly in the south and east where there are more large trees. Later on in the year they disperse and may be seen out in the open. Recently, while on coastwatch duty at Gwennap Head lookout, a squirrel joined me on the balcony and proceeded to sun itself. This is one of the most exposed headlands in Penwith! They were originally from North America and were introduced into England and Wales from 1876 onwards.

Mice, voles and shrews are numerous in and around the woods and in the countryside in general. Little work has been done on this group and the status of the different species is unclear. However, a species list is given in the appendices.

* * *

WOODLAND
Butterflies

Unlike most British butterflies, the Speckled Wood lives in partially shaded conditions and so it is common in woodlands. The male adopts a patch of sunlight as its territory and in this it basks, displays, chases off male intruders and pursues females of its own kind. This behaviour is fascinating to watch because few butterflies are very territorial. The dark brown wings speckled with cream blend perfectly into the background pattern of dappled sunshine in the woodland clearings. Eggs are laid on selected grass blades on which the caterpillars will feed, and give rise to three broods of adults flying between April and September. The colours of the last brood are paler. It is a common butterfly today, but it was scarce in the nineteenth century: shadier conditions resulting from a lack of management in British woodlands have most probably led to this increase.

High summer is the time when the large and magnificent Silver-washed Fritillary is flying, although there are only a few old records for the extreme west of Cornwall. It is found only in a few localities that have mature, deciduous woodland. Any large, orange butterfly flying high in the treetops is more likely to be the Dark Green Fritillary, which also inhabits woodland rides and clearings, but even this species has become quite rare. The orange on the upperwings of both species is marked with black dots and lines, but the presence of four silver streaks on the underside of the greenish hindwing is diagnostic of the slightly larger Silver-washed Fritillary. The caterpillar feeds on Dog-violet leaves and the eggs are laid on a tree close to this food plant. Comma butterflies, also orange-brown, tend to be found on woodland edges or in the vicinity of tall hedges.

Holly Blue butterflies fly several feet up, around the foliage of trees and bushes, particularly holly, ivy and gorse on which it lays its eggs. They are by no means confined to this habitat, being common in the countryside generally and also in suburban gardens. The silver blue of the underwing is marked very simply with a few tiny, black dots, with no orange coloration as there is in the Common Blue. The population is subject to marked fluctuations and recent research has shown that this is linked with a parasite of the larval stage. Some years it is common, with two broods emerging between March and September.

Dragonflies

One seldom gets more than a glimpse of the Southern Hawker *Aeshna cyanea*, which is a large dragonfly that has obvious associations with water, but which may also be seen in woodland rides or along hedges. On fine days the dragonfly may be seen flying fast and erratically around the trees, usually pitching high up and out of view. Even when it settles at eye level it is quick to detect movement and take flight. The Common Hawker *Aeshna juncea* and the Migrant Hawker *Aeshna mixta* are also recorded locally, although the latter is not usually found in woodland. Specific identification of these three hawkers is very difficult unless close views can be obtained.

6

FARMLAND

Until the mines began to fail the Cornish people had neglected husbandry preferring the chances of quickly gained wealth from the rich minerals scattered below the surface of the earth to the increasingly monotonous toil and moil of ploughing and reaping.

Richard Carew (1602)

It does seem that farming was once the Cinderella of the Cornish economic order, but past or present, thriving or not, it had and still has a fundamental effect on the landscape and the wildlife it supports.

Intensification and mechanization took a long time to reach this part of Cornwall, and it is only relatively recently that farmers began to grub out hedges on a large scale for the convenience of modern machinery and to gain a little extra ground. Many fields became grossly enlarged and sometimes the removal of hedges altered the drainage system, causing silt run-off and flooding away from the fields. Special skills and a great deal of time are involved in the construction of a Cornish hedge, but it can be ripped out in an instant by one man operating a machine. Often little or no thought is given to why it was built in the first place. Now at least some of the remaining hedges have legal protection. Before this, landscapes such as this one described by Hudson, were under threat: 'Now after centuries of this process of removing and piling up stones, the farmland has become covered over with a network of enduring hedges, or fences, intersecting each other at all angles; and viewed from a hill-top, the country has the appearance of a patched quilt made of pieces of all sizes and every possible shape, and of all shades of green from darkest gorse to the delicate and vivid green of the young winter grass.' The landscape described was most probably in the Zennor area and would be much the same today, as most of the intensification took place in the south of the peninsula. The rocky and uneven nature of the northern terrain makes reclamation a costly and difficult business and, moreover, a large proportion of this moorland has ESA status (Environmentally Sensitive Area) and such a practice would be disallowed.

The overall pattern of farming in Britain has altered, with mixed and small-scale farming on a local subsistence level being replaced by the large-scale culture of whatever product happens to do well in a particular district. Export of most of the produce follows. So it is that the main crops grown in the extreme west of Cornwall have been potatoes, turnips, broccoli and grass; this last for beef production and dairy cows. All of these have in some way been affected by the current crisis in farming, particularly with dairy and beef cattle, many farmers selling up or looking for alternative incomes. On a smaller scale, cereals and flowers (daffodils, kaffir lilies, violets and anemones) are grown.

Grass fields and pasture bear no resemblance to natural meadows; most have been ploughed and re-sown with cultivated varieties of high-yield grasses, usually Italian Rye-grass *Lolium multiflorum* and, nowadays, two or three cuts of silage are taken off rather than one cut of hay. Some cattle, mostly beef, are put out to graze in the fields, and sheep are often pastured in the winter to keep the land in good heart. Borlase wrote in 1758 'The sheep of Cornwall in ancient times were remarkably small and their fleeces so coarse that their wool bore no better title than that of Cornish hair, and under that name the cloth made of that wool was allowed to be exported without being subject to the customary duty paid for woollen cloth'. He also makes a reference to the small sheep of Sennen, saying that they gave sweet mutton due to an abundance of snails in the dune grasslands, which were supposed to 'yield fattening nourishment to the sheep'. In 1907 Hudson wrote that pigs ranked second to the small Jersey-like cow but, at the same time, he remarked on the poor condition of the cows in Penzance market.

The trend towards intensification meant that it was no longer economically viable to cultivate the small fields on the sloping cliffs or to use unbroken land for grazing. These changes have affected animals and plants alike, some of which have been discussed in the preceding chapters.

The lack of grazing on unbroken land within a farm or where common rights were held has affected some butterflies because the caterpillars of several species feed on low growing plants, and low intensity grazing kept back competitive grasses, etc. This was exacerbated by myxomatosis in the rabbit population. Examples are the fritillaries and Small Coppers which feed on violets and sorrel respectively. Common Blues were affected by the loss of natural grassland through intensification, and also by the shading out of their food plant, Bird's-foot-trefoil. Intensification, including hedgerow removal, has also affected the brown butterflies that were once so common in hay meadows, which have now been replaced with silage.

Plants

A lot of the land in this peninsula that is now under cultivation would originally have been heathland or low scrub. Extensive natural grassland would only have existed on cliff slopes and in moorland bogs. When land was first reclaimed and sown with native meadow grasses, with no herbicides or chemical fertilizers added, a herb-rich sward would have been created, although this was obviously less productive than today's monocultures. Maximum production has only been made possible by sacrificing food quality for quantity.

The early grass and cereal crops would have been interspersed with attractive 'weeds', some of which still grow in odd field corners and in the hedges. They include species like the Common Knapweed *Centaurea nigra*, Pale Flax *Linum bienne*, Yellow-rattle *Rhinanthus minor*, Corn Marigold *Chrysanthemum segetum* and many other common ones. Corn Marigold grows well on our soil and is sometimes common on field edges, where it has escaped treatment with lime. I have seen neglected fields of golden barley mixed with Corn Marigold and Scentless Mayweed *Tripleurospermum inodorum*, and that was a wonderful sight. There were also some grass fields near the north coast in which hundreds of flowering spikes of the Southern Marsh-orchid and dozens of plants of Yellow Bartsia *Parentucellia viscosa* grew.

Some seeds can lie dormant for years, only germinating when the

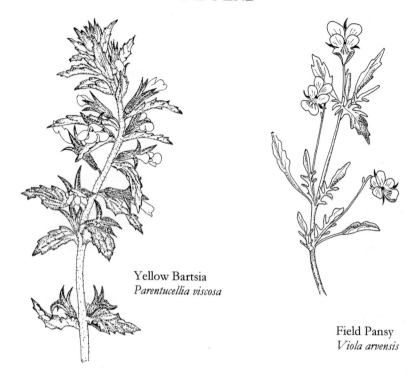

Yellow Bartsia
Parentucellia viscosa

Field Pansy
Viola arvensis

time is right for them. Locally, plants like Common Knapweed, Yellow-rattle and Self-heal *Prunella vulgaris* may appear in grass fields that have fallen into disuse, as long as there is not a heavy residue of chemicals.

Inevitably, when fields are left fallow for a while, weeds grow and, among many coarse looking ones like docks and oraches are some prettier ones. The common Field Pansy *Viola arvensis*, for example, has yellow and white flowers that are just like miniature pansies, while its relative Wild Pansy *Viola tricolor ssp. tricolor*, which has only been recorded in one locality on the peninsula, has the most beautiful flowers varying in colour between mauve, yellow and white. The incredibly tiny, blue flowers of the Cornsalad or Lamb's Lettuce *Valerianella sps.* are clustered together in small umbels surrounded by a ruff of bracts. Corn Spurrey *Spergularia arvensis* is a straggling plant, with leafy tufts springing from the axils that give it a whorled appearance. Its flowers are small and white. Scarlet Pimpernel *Anagallis arvensis* ssp. *arvensis* is familiar to most people with gardens, but few have seen the lovely, rich blue variety of this flower, which occurs locally in the district. Speedwells are quite as lovely as many garden flowers, especially the deep blue Germander Speedwell *Veronica chamaedrys* and the introduced Persican Speedwell *Veronica persica* where the blue fades into chalky-white on the lower petal. Both these 'weeds' are common, but less so is the Lesser Snapdragon or Weasel's-snout *Misopates orontium*, the bright, rose-pink flowers of which may appear suddenly in gardens. Hairy Bird's-foot-trefoil *Lotus subbiflorus* has a restricted distribution in Britain and is considered a rather scarce, mostly coastal species in West Cornwall, but it can occur as a common weed in fallow fields, sometimes with quantities of Trailing St John's-wort *Hypericum humifusum*.

One of the greatest assets to the farming landscape is the Cornish hedge, traditionally built of stone with an earth infill and, as described by Hudson, 'man's creation, but nature has made them what they are'. They have an important role in that they act as corridors between different wildlife habitats and they are also ideal refuges themselves for both plants and animals in what may be acres of sterile farmland. One of the great delights of spring in Cornwall is to watch the hedges come into bloom and then change colour as certain flowers die and others take their place. First the hedges are a glorious mixture of blue, pink and white with the flowers of Bluebells, Red Campion and Greater Stitchwort *Stellaria holostea*, interspersed with lace-like Cow Parsley *Anthriscus sylvestris* flowers. Shortly after, there is a magnificent show of Foxgloves, with an occasional pale pink or white bloom adding an element of curiosity. When these early flowers die back, they are replaced by bright yellow hawkweeds *Leontodon sps.* and cat's-ears *Hypochaeris sps.* complemented beautifully by deep pink Betony *Stachys officinalis* and mauve Sheepsbit flowers.

Often these days the hedges are cut between these two different phases and, for a short while in late July, the hedges are brown and dull. Several other common flowers find a niche between the dominant ones or put on a bit of a show themselves, like the English Stonecrop, Creeping Cinquefoil *Potentilla reptans*, Lesser Stitchwort *Stellaria graminea*, Golden-rod *Solidago virgaurea*, Yarrow *Achillea millefolium*, Meadow Vetchling *Lathyrus pratensis* and Restharrow *Ononis repens*. The pale lilac

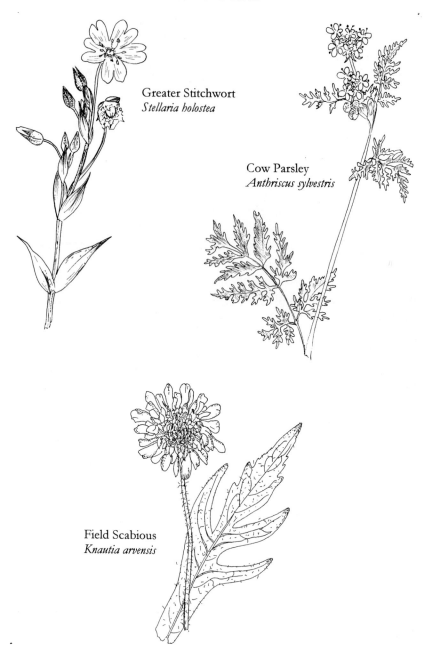

Greater Stitchwort
Stellaria holostea

Cow Parsley
Anthriscus sylvestris

Field Scabious
Knautia arvensis

FARMLAND

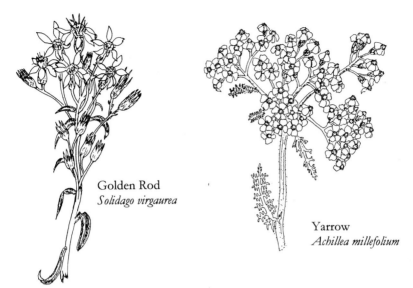

Golden Rod
Solidago virgaurea

Yarrow
Achillea millefolium

flowers of Field Scabious *Knautia arvensis* appear sparingly in some of the hedges, but it is by no means a common plant. The flowers come out in late summer and are quite conspicuous on their very long stems. Old hedges that have not been cut regularly may become overgrown with gorse and blackthorn and, in April, burst into yellow and white blossom. Hedges may also support small trees, usually elder, ash or sycamore, the roots of which grow into it with such vigour that they become an integral part of it.

Sometimes these hedges provide information about the past history of the location. For example, if there are lots of bluebells, wood sorrel and primroses in the hedge, it is likely that the site was once wooded, just as a predominance of heather, gorse and bilberry implies that it used to be heathland.

Ancient trackways, kept open by tractors for access to fields, can be good hunting grounds for plant enthusiasts, especially in wet ground near moorland. In areas of shallow water the spring-flowering water crowfoots send up buttercup-like flowers with white petals and yellow centres. The Ivy-leaved Crowfoot *Ranunculus hederaceus* is one of the two common species and is distinguished by the sepals and petals being of equal length, whereas the petals of the Round-leaved Crowfoot *Ranunculus omiophyllus* are twice as long as the sepals. The first indication of Chamomile *Chamaemelium nobile* is usually the sweet, heady smell emitted when its leaves are crushed underfoot, and then the feathery leaves and the flowers, with their unusual, backward-deflecting petals, are soon noticed.

Water Purslane *Lythrum portula* and Blinks *Montia fontana* are common plants in these trackways, but the delicate strands of Coral Necklace *Illecebrum verticillatum* are a local rarity. This plant has whorls of tiny, white flowers on red stems. It has been recorded on a farm track near Porthmeor, where it grew with another uncommon plant, the Chaffweed *Anagallis minima*, which has minute, pink flowers and is one of Europe's smallest plants.

Chamomile
Chamaemelum nobile

Birds

The Corncrake has disappeared as a breeding bird from Cornwall since mechanization and intensification. According to Rodd (1880), it was not uncommon in grass fields and cloverseed in the county generally; in 1915 it was 'sparingly distributed' in the Land's End area (A.W.H. Harvey) and, by 1931, G.H. Harvey was noting an absence of its unmistakable call in some summers in the Penzance district. The sound is not unlike that produced by running one's finger along a comb and usually this was the only evidence to be had of the presence of this bird. It cannot be described as musical but it can be an evocative sound for those who have heard it calling in Scotland as the evening light falls on its lovely island refuges. When cereals and grass were cut by hand scything, the nesting birds had a chance to escape but, as soon as machines took over, adults, eggs and young alike were destroyed. Nowadays, cutting takes place earlier, when birds are most vulnerable, say in June as opposed to mid-July, which gives them no chance at all.

Farmers in the north of Britain with Corncrakes breeding on their land are asked to cut from the centre of the field outwards to allow the birds to get away. Several are cooperating and are even proud of 'their birds'. Except as a rare migrant, the Corncrake is no longer with us in the south-west. The last breeding record in Cornwall was in 1962.

The Corn Bunting spread across Europe with the creation of the agricultural landscape and was once fairly common in this peninsula. Sadly, it is now extinct here. The musical jingle of notes resembles the sound of tinkling glass and was a very familiar sound to Hudson in 1908. Indeed, the bird was considered by most other ornithologists of the late nineteenth and early twentieth century to be fairly common here, but was confined to open, treeless country, and a study by Ryves and Ryves in 1934 showed that most birds were nesting in gorse and bramble. The decline of this species generally, even since the early 1970s, has become obvious and the reasons are most likely to be related to the intensification of farming, particularly the lack of winter stubble and weedseed. So now the jingling song of this rather plain bunting, which Hudson associated with 'green or yellow fields and sultry weather', is heard no more.

The Cirl Bunting is another species which has been declining in Britain, and has only a tenuous foothold in Cornwall. There are no specific records of breeding birds in the peninsula, and Penhallurick (1978) concludes that it was never common west of Penzance. Once again intensification is considered to be one of the reasons for its decline, but cold winters may also be a factor because it is on the northern edge of its range, meaning that conditions here are less than optimal for its survival.

Although there are many tales of decline and extinction, there are still some birds to be seen in the agricultural zone, but they are more particularly associated with marginal land and other features of the farming landscape like hedges, wet valley bottoms and old barns. Set-aside land has advantages in providing weedseed and cover for breeding, but only if it is not subjected to frequent spraying and early cutting.

Yellowhammers have a preference for marginal land where farmland butts onto moorland or rough ground. The brilliance of the yellow plumage, particularly in the male, can come as something of a surprise; it can be as bright as the yellow of the gorse flowers with which it is

Yellowhammer

associated. It must have been more numerous in the peninsula at one time because Hudson recalls flocks of seventy to eighty birds sheltering around the farms, and he gives a classical description of this: 'I was standing out of doors, when the sun came out beneath a bank of dark cloud and shone level on the slate roof of a cow house. It was an old roof on which the oxidised slate has taken a soft blue-grey or dove colour – the one beautiful colour seen in weathered slate; and no sooner had the light fallen on it than a number of Yellowhammers flew from some other point where they had been sitting and dropped down upon this roof. They were scattered over the slates and sitting motionless with heads drawn in and plumage bunched out, they were like golden images of birds, as if the sun had poured a golden-coloured light into their loose feathers to make them shine.'

Greenfinches used to be common around farmyards, feeding on spilt grain and chaff, but with changes in the type of produce and modern methods of harvesting and storage, this food is no longer available to them in useful quantities. Dressed grain has been cited as another reason for a decline in their numbers, but there seems to be no clear evidence for this. Instead, the birds have acquired a suburban habit. They are now more frequently found in gardens, churchyards and other well-vegetated places associated with human habitation.

The current trend towards barn renovation for holiday homes in Cornwall has had an effect on the Barn Owl population. Many old granite barns attached to farms have been enlarged and renovated (not always tastefully). Some of them were traditional nesting sites for

FARMLAND

Clara Vyvyan's 'Holy Athenian Bird', which had gone unmolested for so many years. It has alternative nesting sites in trees, man-made stone embankments and quarries, but still the population is small and insecure. In 1999, public sympathy for a pair of breeding Barn Owls held up roadworks on the Catchall to St Buryan road for many months. The birds were breeding in an embankment which had partially collapsed. Unfortunately, they have not taken to the nest box provided for them and have since disappeared. Dawn and dusk are the best times to see this most beautiful, white owl as it banks and glides along hedgebanks and over rough ground.

Swallows are also affected by the renovation of barns and old cottages. Ornithologists believe that before they took advantage of

Swallow

these man-made sites they nested on the coast in caves and fissures. A small population breeds locally on houses and barns, but greater numbers are seen during their migration, especially in autumn, when adults and young from near and far gather together socially on the telegraph wires before they undertake the hazardous journey across the English Channel. Every autumn there are large gatherings around Porthgwarra, which acts as a last staging post.

Magpies have increased since the days when they were seriously persecuted by gamekeepers and now they are very common, especially where there is a mixture of farmland and rough ground. While on his lengthy rambles in the peninsula, Hudson became particularly well acquainted with this bird of character: 'You meet with him twenty times a day when out walking. He flies up at a distance ahead, rising vertically, and hovers a moment to get a good look at you, then hastens away on rapidly beating wings and slopes off into the furze bushes, displaying his open graduated tail.' Hudson was aware of the status of the Magpie in other parts of the country and considered that magpies were always common here because of the lack of pheasants and therefore of gamekeepers! Nowadays, the loud chatter of magpies is certainly one of the commonest rural sounds locally. It is often disliked because it preys on the eggs and young of small birds, but there is ornithological evidence that this is not contributing towards the decline in numbers of small bird species and is simply part of the normal predator/prey relationship that serves to ensure the fitness and survival capacity of a population. Indeed, most authorities believe that the increase in the number of domestic cats is a much greater threat.

The pheasants that we see in the countryside today are various mixes of the Black-necked race *Phasianus colchicus colchicus*, the Chinese Ring-necked *Phasianus colchicus torquatus* and the Japanese *Phasianus colchicus versicolor*. It is thought to have been an English introduction to Cornwall that had established a feral stock by the late sixteenth century. However, many of the pheasants in this peninsula have been raised and released by farmers for winter shooting, although feral birds do occasionally breed successfully. Partridges are also released, and these may be Chukars, Red-legged Partridges or a bird that closely resembles our native Grey Partridge. This latter species has been declining nationally for a long time, but the decline has accelerated in the last two decades. It is now certainly extinct in the peninsula for many reasons, mostly to do with changes in farming practices. Trapping and lamping for rabbits is also thought to have affected the population considerably in Cornwall.

Arable farming always conjures up pictures of gulls flying in the wake of the plough and, indeed, this is as familiar a sight today as it was in the early 1900s when ploughs were drawn by horses rather than tractors. It was this very scene in another part of Cornwall that inspired Daphne du Maurier's story *The Birds*. The gulls are taking advantage of the invertebrates which are turned up in the soil and they are often joined by flocks of corvids (Rooks, Crows and Jackdaws).

Large flocks of corvids are capable of causing extensive damage in cereal fields when they descend on the crop and flatten it. Rooks, often accompanied by their newly fledged young, visit newly sown fields and appear to be stealing the sprouted grain, although it is possible that it may also be leatherjackets that they are after. Even when watching them closely through binoculars, it is difficult to establish what they are taking.

Pigeons, wagtails, pipits and finches spend a lot of time foraging in stubble fields for their individual prey items and so these are excellent places for birdwatching, but stubble fields are virtually non existent these days - a great shame for the birds and the birdwatchers!

Groups of Redwing, Fieldfare, Starlings, Curlew and Common Snipe feed and roost in cultivated or grass fields. Lapwing and Golden Plover appear to be very selective in their choice of fields for feeding, probably because some fields are more productive than others for certain prey items for reasons that are not clear, but which are mostly likely to do with the amount of spraying with pesticides. On bright midwinter's days, it is a joy to see hundreds of Golden Plover and Lapwing in the air and particularly to hear the flute-like calls of the former. There is usually a large mixed flock in the fields around Sennen Village in the winter months, although flock sizes have diminished over the years.

Whimbrels may join flocks of Curlew in the fields in order to rest and feed during their spring and autumn migration. They are similar, but slightly smaller, and are more elegant with shorter bills and a pale crown stripe. Their call is a repeated whistle, eagerly listened out for by bird watchers in the early spring.

Butterflies

Cornish hedges, colonized by a great variety of herbs and shrubs, naturally attract many butterflies. Gatekeepers abound in good years and are noted for their lively, skipping flight. Like many butterflies, they sip nectar and this species seems to have a preference for yellow flowers, although they are also greatly attracted to bramble. Generally speaking, they are found where fine-leaved, wild grasses and shrubs grow together on the sunny side of a hedge. The upperwings of the butterfly are marked with a black eyespot and are orange with a grey border that matches the grey, upper side of the hind wings which also have an orange band. The eyespot is also visible on the underside of the upperwing. This butterfly and the larger Meadow Brown are easily confused, but one difference is that the eyespot of the former has two tiny white pupils, while that of the latter has only one.

A Comma butterfly appears bright tawny orange in flight and is then very conspicuous but, when it pitches and closes its wings, it resembles a dried, shrivelled leaf. This excellent camouflage is brought about by the jagged edges of the wings aided by the cryptic coloration of the hindwings. Close-up, the name-giving white comma on the underside of the hindwing is clearly visible. Commas are recent colonizers of Cornwall, reaching West Penwith about 1939, but they are scarce here. They are found around woods, in shrubby gardens and along hedgerows, showing a preference for full and sheltered hedges. Elm, hop or nettle leaves are selected for egg-laying.

Green-veined Whites like damp places and may be seen patrolling lush hedgebanks in which their food plants, a variety of crucifers, grow; among them are Garlic Mustard *Alliaria petiolata* and Hedge Garlic *Sisymbrium officinale*. These butterflies are similar to Small Whites, but they have prominent grey-green bands along the veins of the underwings, although this is rather faint in the second brood. In the same family, the Orange Tip butterfly also frequents hedges on farmland. The female looks very like the Small or Green-veined White in flight but, when it settles, the distinguishing mottled green underside of the hindwings is apparent. Only the male has bright orange wing-tips. These two white butterflies herald the arrival of good weather in spring, both emerging around mid to late April at roughly the same time as the Speckled Wood. Their relatives, the Small and Large Whites, have very widespread distributions as any cabbage-growing gardener will know, and they visit wild, hedgerow plants of the crucifer family on which to lay their eggs. Immigrant butterflies from the continent boost our Large White population in the summer and there may also be emigration the other way.

7

TOWNS AND VILLAGES

The rural appearance of the scene is enhanced by the adjacent cottages also dressed in ivy; the smoke from the chimneys appearing to rise from masses of green leaves, roses and creepers hang all about the little windows and the flowers in the garden grow almost luxuriantly.

J. T. Blight (1861)

It would be hard to banish nature completely from our doorstep and neither would the majority of people want to. Except in the most urban of environments, there are always some resilient plants and animals which find a way to live within our crowded communities. Towns and villages are made more attractive when nature arrives uninvited and spontaneous; wild flowers and ferns colonize walls, birdsong issues from gardens and butterflies visit cultivated flowers. It would be dull without them and an injustice to alienate ourselves from our natural compatriots.

Plants

Towns and villages within the peninsula are small enough to contain much to interest the natural historian. The plants especially are well worth studying because there are many enthusiastic gardeners here experimenting with plants from warmer countries, even from sub-tropical ones. Extremely cold temperatures and hard frosts are rare, so many of them can survive well. Town gardens, parks and estate grounds are planted with an interesting variety of such plants. For example, it is possible to wander among tall tree-ferns in Trengwainton Gardens or sit among palms, giant echiums and agaves in the town parks and imagine yourself in some distant, foreign country. Penlee Park, Morrab Gardens and Trengwainton Gardens have fine collections of camellias, magnolias, rhododendrons and azaleas and one may also come across some unusual trees like a Judas Tree from the eastern Mediterranean, or a Bladdernut or Yellow Kowhai from New Zealand.

Bear's-breech *Acanthus mollis* and Agapanthas grow very successfully and the former has naturalised itself on roadsides at Mousehole and at Trewidden. It is not easy to grow Banana Palms outside, but a garden at Buryas Bridge had three trees, which somehow survived the hard winter of 1986-'87. The extremely low temperatures experienced during this winter (-10° centigrade) took their toll and, in the town gardens, palms and myrtles were especially badly affected, if not killed outright.

Out of town, the National Trust gardens at Trengwainton, near Madron, are open to the public, and the privately owned gardens at Trewidden to the west of Penzance are open on certain days in the year. There are also some large, private gardens, for example at Penberth and Cot Valley, where many interesting plants can be seen from outside the garden boundary. Small gardens should not be overlooked, because they may too contain unusual plants.

Alexanders
Smyrnium olusatrum

It was gardeners who were unwittingly responsible for the original introduction and cultivation of some of the undesirable escapes, like the Japanese Knotweed, Hottentot Fig and Three-cornered Garlic, as well as many others that are not so aggressive. Many became established directly from the dumping of garden waste, examples being Montbretia and the variegated Yellow Archangel *Lamiastrum galeobdolon ssp. argentatum*. The Three-cornered Garlic is now well naturalised on urban roadsides, as is the common umbellifer, Alexanders *Smyrnium olusatrum*.

This latter is an old introduction from the Mediterranean, which was once grown as a vegetable in the same manner as celery.

Winter Heliotrope or Jack-by-the-hedge *Petasites fragrans* is renowned for being both tenacious and persistent. It carpets many roadside verges, especially in the east of the peninsula, and its pugnacious shoots have even been seen breaking through thin tarmac! With equal persistence, Clara Vyvyan struggled to rid her garden of it: 'I begin to think I shall have to follow it sideways across the sea to America or downwards to Australia, but the brittle roots will always snap before I get near those continents'. Not everyone dislikes it though: the strongly vanilla-scented flowers are a welcome sight in a bleak and grey December standing aloof from their green carpet of large, disc-shaped leaves. Hudson wrote of a strange sight he came upon in Madron churchyard, 'when the great frost killed the mass of vivid green leaves, leaving them brown and flat on the earth, exposing the upright flowering stems to full advantage.'

There are many other garden escapes in towns and villages, some of which are very localised like the Green Alkanet *Pentaglottis sempervirens*, which has the most beautiful azure-blue flowers, and the Pencilled Cranesbill *Geranium versicolor* that has a delicate tracing of dark veins on its pale pink petals. Cultivars of Yellow and Pink Oxalis tumble over granite walls; the three-lobed leaves and the brightly coloured flowers are superficially similar to those of its wild relative, the Wood-sorrel. Biting Stonecrop *Sedum acre* is a native plant that forms tight cushions on walls and even on the slate roofs of some of the charming old cottages in Mousehole, where it is said to bring good fortune by protecting the household from fire. This species differs from English Stonecrop in its yellow flowers and also in that its leaves are hot to the taste, like mustard, which presumably is why it is 'biting'. Snow-in-summer *Cerastium tomentosum*, a recent garden escape, is aptly named because its pure white flowers spring from a mass of silver-green foliage and, with a bit of poetic license, resemble a new fall of snow. A granite wall provides the perfect setting for this plant.

Damp places on shaded walls or the bare, stony corners of buildings are ideal situations to look for the bright green, mossy growth of Mother-of-thousands or Mind-your-own-business *Soleirolia soleirolii*. This is another well naturalised introduction and was first recorded in

1917. The trailing stems are thickly endowed with tiny leaves which mostly conceal the minute, solitary pink flowers. It is now extremely common around habitation. Another very common plant that has naturalised itself on stone walls in towns and villages is the Ivy-leaved Toadflax *Cymbalaria muralis*. Its flowers are like small, mauve and yellow snapdragons. A variety with pale cream flowers and light green leaves occurs around Penzance. Clusters of deep mauve, bell-shaped flowers of a cultivated campanula, which resembles the Nettle-leaved Bellflower, appear on urban stone walls. This has spread from gardens and seems to survive very well even when there is no soil at all. It looks wonderful against the bare stonework.

Cornsalad has been noted already as an arable weed, but clumps of it will also grow on stone walls in town. Red Valerian *Centranthus ruber* is a popular garden plant that is common on garden walls, seeding itself readily so that it spreads quickly. The flowers come in many shades of red, pink and white and, although they have an acrid smell, they are particularly attractive to the colourful butterflies, i.e. Red Admirals, Peacocks, Small Tortoiseshells and Painted Ladies. Buddleia *Buddleja davidii* has naturalised itself on some derelict sites around the peninsula, usually in urban or suburban areas. It also has a special attraction for butterflies and is sometimes called the Butterfly Bush.

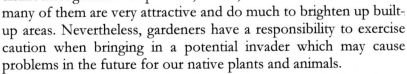

Red Valerian
Centranthus ruber

It is often considered puritanical to dislike these garden overspills and, indeed, many of them are very attractive and do much to brighten up built-up areas. Nevertheless, gardeners have a responsibility to exercise caution when bringing in a potential invader which may cause problems in the future for our native plants and animals.

There is no dearth of native plants in the towns and villages here.

Not so many years ago, when local people used wild plants for medicinal purposes, a favourite, general remedy was made from an infusion of Pellitory-of-the-wall *Parietaria judaica*. This grows commonly on garden walls or buildings and is not at all striking in appearance, being small and bushy with tiny, green flowers that form clusters up the red stems.

Spurreys are small and fleshy with short, needle-like leaves that are slightly swollen. There are three common species that grow here, including the Corn Spurrey that has already been cited as a field weed. Rock Spurrey *Spergularia rupicola* and Sand Spurrey *Spergularia rubra* both have pink flowers with yellow centres, the latter distinguished by the petals being shorter than the sepals, while they are of equal length in the former species. Sand Spurrey is found in dry habitats in towns and in the countryside, but the Rock Spurrey is a coastal species.

Red Admiral

Before finishing this account of the flowering plants of the peninsula, mention should be made of St Just Airfield, which could be considered a suburban habitat. In late summer the turf away from the runways, which is left un-mown for a while, is an extremely rich community, having had no chemicals applied to it. A list of plants would include Wild Thyme, Self-heal, Devil's bit Scabious, Birds-foot-trefoil, Hawkbit, Eyebright, Red Clover, and best of all Autumn Lady's Tresses *Spiranthes spirales*, an orchid of which there were thousands in August-September 2000.

Small Tortoiseshell

TOWNS AND VILLAGES

Ferns lack the array of bright colours found in flowering plants, but they compensate for this by their elegant and usually symmetrical growth form. The fronds can be simple or intricately divided, and mostly spread out from a central rootstock. Some of the smallest occupy tiny cracks and crevices in walls, breaking up the monotony of the stone. A common species in towns and villages is the very delicate Maidenhair Spleenwort *Asplenium trichomanes*, which has dark green fronds divided into rounded pinnae that are attached to black, thread-like stems. Another is the Wall-rue *Asplenium ruta-muraria* that grows in small tufts and is a dark olive green, except for the base of the stem which is darker. Maidenhair Ferns are popular house plants because of their lovely, cascading foliage, and this must account for its appearance in some of the gardens in Penzance. Otherwise, it is an extremely rare native of damp cliff-faces and occurs at Lelant on the north coast, just outside the area described in this book.

Within the peninsula are many old granite cottages and farmhouses that are very much in keeping with the surrounding countryside. This is because they are built of the local stone and have natural slate roofs. They are, in Hudson's words, 'weathered and coloured by the wind, sun and rain, and many lowly vegetable forms to a harmony with nature'. Some of these lowly vegetable forms are lichens, which grow on building stone and slate in the same way as they do on the natural rocks. The most common one is bright yellow and belongs to the family Xanthoria.

Birds

Suburban gardens, especially large ones with a good variety of shrubs and trees, are excellent habitat for the commoner bird species. Blackbirds, Robins, Dunnocks and Wrens take up territory in spring and cheer the neighbourhood with their singing, but a quiet period follows when they are rearing their young. However, autumn in the south-west brings an influx of robins from the north and east, which compete aggressively for winter territories, when both male and females begin to sing again. Also, young birds of most species experiment in the late summer, copying and learning from the adults to produce a subdued and shortened version of their normal song (the sub-song), which is a run-up for the real event the following year.

Food scraps and specialised bird-feeders supplement the diet of these birds and increase their chances of survival over winter. Nut baskets, seedcake and rotting fruit attract other species like tits, finches, thrushes and even woodpeckers to gardens, bringing much pleasure to the human occupants. Unfortunately, these days, large numbers of domestic cats often clean out populations of garden birds.

Around most villages and towns there are small concentrations of House Sparrows. These birds are always found near habitation and, with their friendly disposition and homely chirruping, they are much loved but often taken for granted. Hudson, always with an eye for beauty, is cruel in his description: 'As for the vulgar sparrow, nothing, not even the miracle working sun, could make him brilliant or beautiful to look at'. The fact that Hudson recorded several hundred issuing from a roost in a disused mining pit and several thousands sitting in a blackthorn bush near St Just supports the opinion that they are nowhere near as common as they used to be. Penhallurick (1978) suggested that changing farming practices, less free-range poultry and lack of nesting sites on modern houses may be some of the factors involved in their decline, an opinion which is generally accepted today.

Brightly coloured Greenfinches and Chaffinches are welcome visitors to most gardens, the former having only recently adapted to life in suburbia. The pretty plumage and the tinkling song of the Goldfinch led to them being trapped and sold as cage birds in the early twentieth century. Even on this remote peninsula, W.M. Harvey noted a heavy toll taken by bird-catchers in 1915. Nowadays the population suffers from the effects of excessive weed control, which denies them sufficient weedseed on which to feed in the autumn and winter. Large gardens often have rough pockets of land where thistles, docks, dandelions and other seed-bearing plants flourish and these are the places to watch out for Goldfinches. Bird lovers usually leave wild areas of land in their gardens, forfeiting tidiness and formality for the pleasure of seeing and enjoying the birds.

The wistful 'pheu pheu' call of the Bullfinch was not always a welcome sound in gardens because, when its natural food was in short

Bullfinch

supply, the bird attacked the newly formed buds of fruit trees. This is how it earned its dialect name of 'bud-picker' (Rodd 1880). In the past, when orchards and market gardens flourished in the Penzance area, it was said by Clark (1906) to have been almost exterminated but in 1915 A.W.H. Harvey wrote that it was holding its own in spite of considerable persecution. The males are handsome with a jet black crown, blue-grey back and vivid rose-pink breast. There is a small population in the peninsula today, which is locally distributed in semi-urbanised valleys and large gardens with well grown scrub, such as are found at Treveal, Porth Curno, and Crean. Generally, they seem to be declining.

No cliff-nesting colonies of House Martins exist on this peninsula; the nearest ones on the Lizard, in the vicinity of Polpeor Cove, are now gone. They nest in small numbers on buildings in towns, villages and on farm houses. Kemyel Drea at Lamorna has a thriving colony (125 nests in 1998) and they are also common migrants, joining in the mass gatherings of hirundines in late summer. Sand Martins differ from House Martins in that they lack a white rump, are brown rather than black above and have a breast band. Old records show that this cliff-nesting species once bred on the road between Penzance and Newlyn (Courtney 1845) but they no longer breed anywhere on the peninsula. There are colonies at Perranuthnoe and Gwithian, east of

Penzance on the south and north coasts respectively. The remarkable agility of martins in flight enables them to snatch insects from the air with great ease and it is a delight to watch them.

The Cornish name for the Swift is 'gwennol dhu' (black swallow), and this was probably also a cliff-nesting species until it began to use man-made structures. Indeed, back in 1931, G. H. Harvey suspected them to be still nesting in cliffs here, but the tradition has certainly died out now. Large numbers of them frequent our towns and villages during the short summer season that they spend with us. The fascination of swifts is that they rarely alight, meaning that they must eat and sleep in the air except when brooding eggs or feeding chicks. They sweep low and fast, screaming as they chase each other through streets and alleys, picking insects from the air. Their activities reach a peak on balmy summer evenings when the young birds are flying with the adults and then, all too suddenly, they are gone south and it is still only the end of July. There appears to be little information available on numbers and breeding sites, but they certainly do breed in churchtowers. The ones at St Buryan and Paul are classic sites.

Jackdaws were commonly known as 'Chawks' in the Land's End district, obviously because of the noise they make. They are well known in the district for their persistence in trying to build nests in chimney pots. They often succeed, too, especially when a chimney is not in use, and many a jackdaw conversation echoes down flues and around kitchens in old cottages. Strategically placed wire-netting seems to be the only answer! They are great characters and full of mischief. Hudson greatly amused himself by imagining what they were saying to each other as they peered down the chimney pots in St Ives. He also remarked on their association with donkeys when removing ticks from the coats of these animals. They do well by associating with man and are extremely common and widespread both in urban and country areas. In the autumn and winter, jackdaws make up part of the large corvid flocks which feed and roost together. Boskenna Woods is the site of a large communal roost.

Like youths on a street corner, gregarious Starlings gather round their nesting sites and chatter ceaselessly. In the early nineteenth century, small numbers were thought to nest around the cliffs in Cornwall and then there was a gradual spread westwards of birds

Starling

nesting in and around habitation. During 1890-'92, they reached Penzance where they apparently caused a nuisance by blocking water pipes in the year 1910. Holes in the stonework of town houses, cottages and barns are popular nesting places, as well as cliff-faces and today there is still quite a healthy population here, although house and barn renovations have ousted many pairs. Starlings take part in cold weather movements and sometimes thousands of them are recorded moving westwards over the peninsula in late autumn or early winter. They may be heading for Ireland, France or Spain. Many of the birds that winter here are thought to originate from Eastern Europe and, rarely, a Rose-coloured Starling will turn up with them, a native of those parts. In a normal winter thousands flock together at the end of the day and then head off to roost communally in certain areas that vary each year; favoured sites are the conifer woods at Pendeen, a fir plantation at Skewjack and the reedbed at Marazion Marsh, just to the east of the peninsula. It is a phenomenon worth witnessing, as flocks spiral overhead like bees swarming in a darkening sky and then, with rushing wings, the birds drop down in such numbers that they blacken the reeds or the trees on which they eventually settle.

8

CONSERVATION

But the sea and its colour and the joy of a vast expanse could not have drawn me so often to the castled forelands nor held me so long but for the birds that haunted them, seeing that this visible world is to me but a sad and empty place without wonderful life and the varied forms of life, which are in harmony with it, and give it a meaning, and a grace and a beauty and splendour not its own.

W. H. Hudson (1908)

This beautiful landscape that has inspired so many writers and artists in their work and which attracts many thousands of visitors each year is in need of sensitive and careful conservation. It is not a prosperous area and the fight by concerned individuals or groups against insensitive development or despoliation is a hard battle to win. Too often, financial incentives are enough to sway the balance (witness the present disruption and development by the various cable companies which threatens to continue because large hand-outs are being offered).

Some of the wild, uncultivated land belongs to various estates, i.e. Bolitho, St Aubyn and Tregothnan and, in most cases, this has afforded it a certain amount of protection. Sometimes simple lack of access or the fact that it may be preserved for shooting or hunting can be beneficial to conservation in a broad sense. Although many of us dislike these sports, they may save the land from being developed or cultivated. The heath above Porthgwarra and the moor below Boswarva Carn, each valuable wildlife habitat, are estate owned.

There is no doubt that if the peninsula was awarded National Park status, as the Pembrokeshire coast has been, matters would be much improved. The regulations are strict even over such small matters as road signs and markings (these latter can turn attractive countryside into a rural suburbia).

CONSERVATION

The Cornwall Wildlife Trust has a presence on the peninsula. It owns land at Bosvenning Common near Newbridge, Baker's Pit by Nancledra, Kemyel Crease wood west of Mousehole and Chun Downs south of Morvah. Housed at its headquarters in Allet near Truro is the County Records Centre, where all significant records of animals and plants in Cornwall are held on a computer database.

English Nature, the government conservation body, owns no land but is very involved in a liaising and advisory capacity, especially where areas which are designated Sites of Special Scientific Interest are concerned. There are many SSSIs, but the two largest here are between Aire Point north of Sennen to Carrick Du west of St Ives and between Porthgwarra and Pordennack Point south of Land's End. These are designated for the importance of their plant communities: coastal heath and coastal grassland.

Much of the moorland between St Ives and St Just, extending northwards to include the cliffs down to high water mark, falls within the boundaries of an ESA (Environmentally Sensitive Area), which gives it added protection. It is an irony that whereas farmers within the area were once paid to break in moorland they now receive special ESA payment to conserve it. Also promoting nature conservation in the area are Countryside Stewardship Schemes, voluntary agreements made with farmers by MAFF (Ministry of Agriculture, Fisheries and Food), whereby they are paid to manage a habitat in a certain way for the benefit of plants and wildlife. These schemes involve such practices as the provision of winter stubble, hedge maintenance and light grazing. A Countryside Stewardship Scheme is currently underway on the moors around Chapel Carn Brea, Bartinney Downs and Caer Bran.

By far the largest conservation organization here is the National Trust which owns, at the time of writing, 3,350 acres, not including Trengwainton. Most of it is coastal or bordering on the coast. Eighteen kilometres of coastline is at present in their ownership. Inland sites include Chapel Carn Brea, Trencrom, Foage Valley and Lanyon Quoit. They also have restricted covenants over a further one thousand, two hundred and seventeen acres. These mean that no changes can take place without the approval of the Trust, although they themselves undertake no management. Examples of covenanted land include Porthgwarra and some of the intricate field systems around Zennor, with their spread of large boulders so enthusiastically described by Hudson.

Whereas English Nature is exclusively concerned with nature conservation, the National Trust has a broader scope of objectives. These are nature conservation, archaeology, access, interpretation and education. As far as the former is concerned, the Trust has a variety of management techniques in practice to conserve the important coastal, moorland and wetland habitats for which it is responsible. If neglected many of these areas would become rank, overgrown and inaccessible and would lose much of the plant and animal interest that is now evident (even so, some conservationists prefer a policy of non-intervention). Grazing, control burning and cutting are the main practices used, often in combination with grazing to follow on from the others and, while some management projects are ongoing and successful, others have yet to be introduced or have not yet been effective. The Trust does not have its own stock, but operates by agreements with stock-owning farmers. Careful control and monitoring of the grazing is required because too much or too little does not achieve the desired effect and can actually be damaging. Examples of grazing being used as a management tool can be seen at Bosigran with Manx Loghtan sheep, Foage Valley where rare native breeds of stock are used and at Treveal cliffs where the Trust have had a long-standing agreement with a local farmer. Also, the Galloway cattle, which graze the cliffs intermittently at Bosigran, no doubt provide very tasty organic beef! Grazing is pending at Carn Galver but cattle must be fenced and a system of cattle grids and low level fencing is being devised so that there is no intrusion on the landscape. All these areas were lightly grazed in the past.

Generally speaking, most people like to see animals in a wild landscape; it represents traditional low intensity grazing and the end product (usually meat) is of better quality.

Apart from broad habitat management, other work which the National Trust is involved in includes the hedging of mine shafts to leave open habitat for bats (especially the rare Greater Horseshoe Bat), the eradication of Japanese Knotweed in Cot Valley, Trencrom and Kenidjack, creating a regime which includes winter stubble to encourage the spread of the rare Purple Bugloss (Echium plantagineum) at Nanquidno, Rhododendron control on the moors, and the upkeep of engine houses and chimneys without damaging any important lichens and mosses. Naturally much monitoring and survey work is involved with all of these projects.

Politically and economically, much has changed on the peninsula in the last few decades, but overall the landscape, particularly the coast, remains unspoilt, give or take a few obtrusive modern dwellings. Much of the past is still evident and it is partly for this reason that I have given many quotations from antiquarian writers. In line with this, I should mention a group of enthusiasts who laid the baseline for the study of local natural history and archaeology. These people were not only responsible for forming the influential group known as the Penzance Natural History and Antiquarian Society, with some very reputable members from all over Britain, but they were also the founders of Penlee House Gallery and Museum, initially known as Penzance Museum. First housed in Market House in Penzance, it was later moved to Penlee House, when the fifteen acre Penlee estate have been purchased by public appeal money. Although exhibiting many artifacts of great fascination and interest to the enthusiast, it was not to everyone's taste, containing, as one visitor commented, 'a mangy lot of skins and skeletons under the charge of an imbecile man'. Some items have gone missing since then, among them 'the pig tail of a Chinaman as he was running away at the siege of Nanking'. One wonders what happened to the rest of him! There are still, of course, many interesting exhibits left to see as well as some high quality works of art in temporary exhibitions or on permanent display.

The Natural History and Antiquarian Society was formed in 1839 and ceased to exist in 1961. For the first fifty years or so, it published annual reports of excellent quality covering a wide range of subjects throughout Cornwall. These can be seen on request at Penlee House, where further details of the society's history can also be found.

My final word is that I hope by writing this book I myself will increase an awareness of the natural riches which the district has to offer to those who have their eyes open and minds receptive to its varying moods.

Appendix

PLANTS OF THE LAND'S END PENINSULA

The following list has been compiled from reference to the *Review of the Cornish Flora*, by Margetts and David (1980), the *Cornish Flora Supplement* 1981-1990 by Margetts and Spurgin, my own records and the recently published *Flora of Cornwall* by Colin French, Rosaline Murphy and Mary Atkinson (1999).

Historically, the standard *Flora of Cornwall* by Frederick Hamilton Davey was published in 1909. The story of this Cornish naturalist is told in a charming book called *Stars in the Grass* by Selina Bates and Keith Spurgin (1994). E. Thurston and C.C. Vigurs published a supplement to the flora in 1922 and they were reissued together in a single volume in 1978. Then followed the works listed above.

The latest *Flora* is an excellent piece of work and is the result of eleven years intensive field survey and research. The authors have produced an Atlas based on tetrad squares i.e. two kilometre squares showing the distribution and abundance of each plant before 1980 and after 1979. For convenience' sake and for ease of future recording, my list has been compiled from the five tetrad squares which cover the area described in this book with the addition of a very small area to the east of Penzance.

Interested botanists should refer to the new *Flora* for a precise definition of boundaries.

My list excludes some plants which are known to have become extinct, and when a plant has not been recorded since 1980 this is indicated by an 0. An introduced plant has the symbol * before it and, if this is questionable, it is preceded by ?. Some introduced plants may be native elsewhere in Britain but not in Cornwall. However they are mostly non-native plants that have naturalized themselves by propogating either by seed or by vegetative means. Sometimes it is uncertain whether planting in a wild situation by man has occurred, in which case this is indicated in brackets afterwards. Others are

Appendix: PLANTS

garden escapes or casuals which do not necessarily become established. Extinctions are still happening and new plants are being added continually, but the list reflects the situation as it was on the publication of the new *Flora*.

Varieties and forms are generally not included in this list, but information about them can be found in the *Flora*. Difficult groups such as brambles and dandelions are listed as aggregates. Much work is still to be done on these groups and botanists should refer to the reference lists for the appropriate identification books.

Nomenclature follows Clive Stace, *Flora of the British Isles*, 1997.

Habitat is coded as follows:-

C = Coast	M = Moorland	Wa = Walls
F = Farmland	P = Pasture	WD = Woodland
G = General	W = Wetland	U = Urban
H = Hedges		

Brackets indicate a tendency to a particular habitat within another.

Status is coded as follows:-

C = Common	Sc = Scarce
FC = Fairly Common	R = Rare
LC = Locally common	

Status follows the Latin name, while habitat follows the English name.

sps. = species   ssp. = subspecies   agg. = aggregate

* *Acanthus mollis* Sc Bear's-breech U
* *Acer campestre* Sc Field Maple WD
* *Acer pseudoplatanus* C Sycamore G(WD)
  Achillea millefolium C Yarrow C H
  Achillea ptarmica FC Sneezewort M W
0* *Aconitum napellus* R Monk's-hood U
  Adiantum capillus-veneris R Maidenhair Fern C Wa
0 *Adoxa moschatellina* R Moschatel WD H
  Aegopodium podagraria FC Ground-elder G
* *Aeonium cuneatum* R Aeonium G(U)
* *Aesculus hippocastanum* FC Horse-chestnut G(H)
0 *Aethusa cynapium* Sc Fool's Parsley F

Agrimonia eupatoria Sc Agrimony G
0 *Agrostemma githago* Corncockle F
Agrostis canina FC Velvet Bent G
Agrostis capillaris C Common Bent G
Agrostis curtisii C Bristle Bent G
Agrostis gigantea R Black Bent F
Agrostis stolonifera C Creeping Bent G
Aira caryophyllea C Silver Hair-grass G
Aira praecox C Early Hair-grass G
Ajuga reptans FC Bugle WD
0 *Alisma plantago-aquatica* R Water-plantain W
Alliaria petiolata Sc Garlic Mustard G(H)
Allium ampeloprasum var. babingtonii LC
   Babington's Leek C H WD
* *Allium carinatum* Sc Keeled Garlic G
* *Allium roseum* Sc Rosy Garlic U
* *Allium subhirsutum* Sc Hairy Garlic G(U)
* *Allium triquetrum* C Three-cornered Garlic C WD H
Allium ursinum LC Ramsons WD
Allium vineale Sc Wild Onion G
Alnus glutinosa FC Alder W
Alopecurus geniculatus FC Marsh Foxtail W
Alopecurus pratensis Sc Meadow Foxtail G(F)
* *Ammi majus* R Bullwort G
Ammophila arenaria LC Marram Grass C
0 *Anacamptis morio* R Green-winged Orchid P (C)
Anacamptis pyramidalis Sc Pyramidal Orchid C
Anagallis arvensis ssp. arvensis C Scarlet Pimpernel G(F) G(F)
0 *Anagallis minima* R Chaffweed G
Anagallis tenella C Bog Pimpernel W
Anchusa arvensis R Bugloss G
0 *Anchusa officinalis* R Alkanet C
Anemone nemorosa LC Wood Anemone WD M
Angelica sylvestris C Wild Angelica W
Anisantha sterilis FC Barren Brome G
Anthemis cotula R Stinking Chamomile G(F)
Anthoxanthemum odoratum C Sweet Vernal Grass G
0 *Anthriscus caucalis* R Bur Chervil C
Anthriscus sylvestris C Cow Parsley G(H)
Anthyllis vulneraria C Kidney Vetch C

Corncockle
Agrostemma githago

Appendix: PLANTS 151

Antirrhinum majus Sc Snapdragon U
Aphanes arvensis FC Parsley-piert G
Aphanes australis FC Slender Parsley-piert M C
Apium graveolens R Wild Celery C
Apium inundatum Sc Lesser Marshwort W
Apium nodiflorum C Fool's Water-cress W
* *Aponogeton distachyos* R Cape-pondweed W
* *Aquilegia vulgaris* R Columbine G
Arabidopsis thaliana R Thale Cress G
0 *Arabis hirsuta* R Hairy Rock-cress C
0 *Arctium lappa* Greater Burdock G. Possible incorrect record.
Arctium minus C Lesser Burdock G
Arenaria serpyllifolia ssp. leptoclados R Slender Sandwort C
Arenaria serpyllifolia ssp. serpyllifolia R Thyme-leaved Sandwort C
Armeria maritima C Thrift C
Arrhenatherum elatius FC Onion Couch G
0 *Artemisia absinthium* R Wormwood H
Artemisia vulgaris C Mugwort G
* *Arum italicum ssp. italicum* C Italian Lords-and-ladies WD U
* *Arum italicum ssp neglectum* R Italian Lords-and-ladies H F
Arum maculatum Sc Lords-and-Ladies WD
0 *Asparagus officinalis ssp. prostratus* R Wild Asparagus C
Asplenium adiantum-nigrum C Black Spleenwort H
Asplenium marinum C Sea Spleenwort C
Asplenium obovatum ssp. lanceolatum Sc Lanceolate Spleenwort H Wa
Asplenium ruta-muraria Sc Wall-rue Wa
Asplenium trichomanes ssp. quadrivalens C Maidenhair Spleenwort Wa
Asplenophyllites × japonica = *Adiantum nigrum × Phyllites scolopendrium* R G
* *Aster lanceolatus* Sc Narrow-leaved Michaelmas-daisy U
* *Aster novi-belgii* Sc Confused Michaelmas-daisy U
* *Aster novi-belgii ssp. belgii* R U
* *Aster × salignus* Sc Common Michaelmas-daisy U
Aster tripolium LC Sea Aster C
* *Aster × versicolor* Sc Late Michaelmas Daisy U
* *Astrantia major* R Astrantia U
Athyrium filix-femina C Lady-fern WD
Atriplex glabriuscula R Babington's Orache C
* *Atriplex halimus* Sc Shrubby Orache C
0 *Atriplex laciniata* R Frosted Orache C
Atriplex patula C Common Orache G(F)

Atriplex prostrata C Spear-leaved Orache G
*　*Avena fatula* Sc Wild Oat G
*　*Aucuba japonica* R Spotted Laurel U
*　*Azolla filiculoides* R Water Fern W
　　　Baldellia ranunculoides R Lesser Water-plantain W
　　　Ballota nigra ssp. meridionalis FC Black Horehound G
0*　*Barbarea intermedia* R Medium-flowered Winter-cress G(F)
0*　*Barbarea verna* C Winter-cress G
　　　Barbarea vulgaris R Winter-cress G
　　　Bellis perennis C Daisy G
*　*Berberis darwinii* R Darwin's Barberry G
*　*Berberis vulgaris* Sc Barberry G
　　　Beta vulgaris ssp. maritima C Sea Beet C
　　　Betula pendula R Silver Birch G
*　*Betula pubescens ssp. pubescens* Sc Downy Birch G
0　*Bidens cernua* R Nodding Bur-marigold W
0　*Bidens tripartita* R Trifid Bur-marigold W
*　*Blechnum cordatum* R Chilean Hard-fern U
　　　Blechnum spicant C Hard-fern WD M
　　　Borago officinalis Sc Borage G(U)
0　*Botrychium lunaria* R Moonwort M P
　　　Brachypodium sylvaticum C False Brome G
*　*Brassica napus* FC Rape F
　　　Brassica nigra C Black Mustard G
　　　Brassica oleracea var. oleracea Sc Wild Cabbage C
0　*Brassica rapa* C Wild Turnip G
*　*Briza maxima* Sc Greater Quaking-grass U
0　*Briza media* R Quaking-grass G
0　*Briza minor* Sc Lesser Quaking-grass G(F)
　　　Bromus hordeaceus ssp. hordeaceus FC Soft-brome G
　　　Bromus hordeaceus ssp. ferronii Sc Least Soft-brome C
　　　Bromus lepidus R Slender Soft-brome F
*　*Buddleja davidii* FC Butterfly-bush U
*　*Bupleurum subovatum* Sc False Thorow-wax U
　　　Cakile maritima Sc Sea Rocket C
*　*Calendula officinalis* Sc Pot Marigold U
0　*Callitriche brutia* R Pendunculate Water-starwort W
0　*Callitriche hamulata* R Intermediate Water-starwort W
　　　Callitriche stagnalis C Common Water-starwort W
　　　Calluna vulgaris C Heather M C

Appendix: PLANTS

* *Caltha palustris* Sc Marsh-marigold W
?* *Calystegia sepium ssp. roseata* Sc Pink Bindweed C
   Calystegia sepium ssp. sepium Sc Hedge Bindweed G(H)
* *Calystegia silvatica* C Large Bindweed G(H)
   Calystegia soldanella R Sea Bindweed C
* *Camelina sativa* R Gold-of-pleasure U
* *Campanula portenschlagiana* R Adria Bellflower U
* *Campanula poscharskyana* R Trailing Bellflower U
* *Cannabis sativa* Sc Hemp G
0* *Capsella batavorum* R U
   Capsella bursa-pastoris C Shepherd's-purse F
* *Capsella patagonica* R U
* *Capsella rubella* R Pink Shepherd's-purse U
   Cardamine flexuosa C Wavy Bitter-cress W
   Cardamine hirsuta C Hairy Bitter-cress G
   Cardamine pratensis C Cuckooflower W
   Carduus nutans Sc Musk Thistle G
   Carduus tenuiflorus Sc Slender Thistle C
   Carex arenaria LC Sand Sedge C
   Carex binervis FC Green-ribbed Sedge M
   Carex caryophyllea C Spring Sedge C
   Carex distans LC Distant Sedge C
   Carex divulsa ssp. divulsa R Grey Sedge M
   Carex echinata FC Star Sedge W
   Carex extensa Sc Long-bracted Sedge C
   Carex flacca C Glaucous Sedge M W
   Carex hirta R Hairy Sedge G
   Carex laevigata R Smooth-stalked Sedge W
   Carex nigra FC Common Sedge W
   Carex otrubae Sc False Fox-sedge W
   Carex ovalis Sc Oval Sedge W
   Carex panicea C Carnation Sedge M W
   Carex paniculata Sc Greater Tussock-sedge W
   Carex pendula C Pendulous Sedge WD
   Carex pilulifera FC Pill Sedge M
   Carex pulicaris Sc Flea Sedge M
0 *Carex punctata* R Dotted Sedge C
   Carex remota R Remote Sedge G
0 *Carex riparia* R Great Pond-sedge W
0 *Carex rostrata* R Bottle Sedge W

    Carex sylvatica LC Wood-sedge WD
    Carex viridula ssp. oedocarpa C Common Yellow-sedge M W
0   *Carlina vulgaris* R Carline Thistle C
0* *Carpinus betulus* R Hornbeam WD
*   *Carpobrotus edulis* LC Hottentot-fig C U
*   *Castanea sativa* FC Sweet Chestnut WD
    Catapodium marinum FC Sea Fern-grass C
    Catapodium rigidum FC Fern-grass C
    Centaurea nigra C Common Knapweed G
0   *Centaurea cyanus* R Cornflower F
0   *Centaurea scabiosa* Sc Greater Knapweed G(C)
    Centaurium erythraea C Common Centuary G
    Centaurium pulchellum R Lesser Centuary C
0   *Centaurium scilloides* R Perennial Centuary G
*   *Centranthus ruber* C Red Valerian U(H WA)
    Cerastium diffusum C Sea Mouse-ear C
    Cerastium fontanum ssp. vulgare C Common Mouse-ear G(F)
    Cerastium glomeratum C Sticky Mouse-ear G
    Cerastium semidecandrum Sc Little Mouse-ear C
0* *Cerastium tomentosum* Sc Snow-in-summer U
    Ceratocapnos claviculata Sc Climbing Corydalis M WD H
*   *Ceratochloa cathartica* R Rescue Brome U
0   *Ceterach officinarum* R Rustyback Fern W
    Chaenorhinum minus Sc Small Toadflax G(F)
    Chaerophyllum temulum Sc Rough Chervil G
    Chamaemelum nobile LC Chamomile M W
    Chamerion angustifolium C Rosebay Willowherb G(M)
    Chelidonium majus FC Greater Celandine U
    Chenopodium album C Fat-hen F
0   *Chenopodium bonus-henricus* R Good-King-Henry G(F)
    Chenopodium murale R Nettle-leaved Goosefoot C
    Chenopodium polyspermum Sc Many-seeded Goosefoot F
    Chenopodium rubrum R Red Goosefoot F
    Chrysanthemum segetum LC Corn Marigold F
    Chrysosplenium oppositifolium C Opposite-leaved Golden-saxifrage W
    Cicendia filiformis R Yellow Centuary W
0* *Cichorium intybus* FC Chicory G
    Circaea lutetiana C Enchanter's-nightshade G(WD)
    Cirsium arvense C Creeping Thistle G
    Cirsium palustre C Marsh Thistle W

Appendix: PLANTS

Cirsium vulgare C Spear Thistle G
* *Claytonia perfoliata* R Spring Beauty G (C)
* *Claytonia sibirica* LC Pink Purslane WD H
Clematis vitalba FC Traveller's-joy G

Traveller's Joy
Clematis vitalba

Clinopodium ascendens R Common Calamint G
0 *Clinopodium vulgare* R Wild Basil G
Cochlearia danica C Danish Scurvygrass C
Cochlearia officinalis C Common Scurvygrass C
Conium maculatum FC Hemlock G
Conopodium majus FC Pignut WD H M
Convolvulus arvensis C Field Bindweed G
* *Cordyline australis* (seedlings) R Cabbage-palm U
0* *Cornus sanguinea* R Dogwood H
* *Coronopus didymus* C Lesser Swine-cress G
Coronopus squamatus C Swine-cress G
0* *Cortaderia selloana* Sc Pampas-grass U
Corylus avellana C Hazel WD H

* *Cotoneaster integrifolius* Sc Small-leaved Cotoneaster G(U)
* *Cotoneaster horizontalis* R Wall Cotoneaster G(U)
* *Cotoneaster simonsii* Sc Himalayan Cotoneaster G(U)
  Crambe maritima R Sea Kale C
* *Crassula helmsii* R New Zealand Pigmyweed W

0* *Crataegus crus-galli* R Cockspurthorn U
  Crataegus monogyna C Hawthorn G
  Crepis capillaris C Smooth Hawk's-beard G
* *Crepis vesicaria ssp taraxacifolia* C Beaked Hawk's-beard G
  Crithmum maritimum C Rock Samphire C
* *Crocosmia × crocosmiiflora* C Montbretia G

0?* *Cruciata laevipes* R Crosswort G
* *Cupressus macrocarpa* FC Monterey Cypress G
  Cuscuta epithymum C Dodder M C
* *Cymbalaria muralis* C Ivy-leaved Toadflax U(Wa)
* *Cynodon dactylon* Sc Bermuda-grass C

0 *Cynoglossum officinale* R Hound's-tongue C
  Cynosurus cristatus Sc Crested Dog's-tail C

0 *Cynosurus echinatus* Sc Rough Dog's-tail C
* *Cyperus eragrostis* Sc Pale Galingale G
  Cyperus longus R Galingale W
* *Cyrtonium falcatum* R House Holly-fern C
  Cytisus scoparius Sc Broom H F
  Dactylis glomerata C Cock's-foot G

0 *Dactylorhiza fuchsii* R Spotted Orchid P H
0 *Dactylorhiza incarnata* R Early Marsh-orchid M (W)
  Dactylorhiza maculata ssp. ericetorum C Heath Spotted-orchid P H
  Dactylorhiza praetermissa C Southern Marsh-orchid G (W)
  Danthonia decumbens C Heath-grass M

0* *Daphne laureola* R Spurge-laurel U
* *Datura stramonium* Sc Thorn-apple F
  Daucus carota ssp. carota C Wild Carrot G

0 *Daucus carota ssp. gummifer* R Sea Carrot C
* *Delairea odorata* Sc German-ivy U
  Deschampsia cespitosa FC Tufted Hair-grass M W

0 *Dianthus armeria* R Deptford Pink C
* *Dicksonia antarctica* R Australian Tree-fern WD
  Digitalis purpurea C Foxglove G
* *Digitaria sanguinalis* Sc Hairy Finger-grass U
  Diplotaxis muralis FC Annual Wall-rocket G

Appendix: PLANTS

Dipsacus fullonum FC Wild Teasel G
0 *Drosera intermedia* R Oblong-leaved Sundew W
Drosera rotundifolia Sc Round-leaved Sundew W
Dryopteris aemula Sc Hay-scented Buckler-fern WD H
Dryopteris affinis C Scaly Male-fern WD H
Dryopteris affinis ssp. affinis Sc WD
Dryopteris affinis ssp. borreri Sc WD
Dryopteris carthusiana R Narrow Buckler-fern M
Dryopteris x complexa = *D. filix-mas x D. affinis* R H
Dryopteris dilitata C Broad Buckler-fern WD
Dryopteris filix-mas C Male-fern WD
* *Duchesnea indica* Sc Yellow-flowered Strawberry G
* *Eccremocarpus scaber* R Chilean Glory-flower U
* *Echinochloa colona* R Shama Millet G
* *Echinochloa crus-galli* R Cockspur G(U)
* *Echium pininana* R Giant Viper's-bugloss U
Echium plantagineum R Purple Viper's-bugloss G
0 *Echium vulgare* R Viper's-bugloss G
Elatine hexandra R Six-stamened Waterwort W
0 *Eleocharis acicularis* R Needle Spike-rush W
Eleocharis multicaulis FC Many-stalked Spike-rush W
Eleocharis palustris ssp. vulgaris C Common Spike-rush W
0 *Eleocharis quinqueflora* R Few-flowered Spike-rush W
Eleogiton fluitans FC Floating Club-rush W
* *Elodea canadensis* Sc Canadian Waterweed W
Elymus caninus R Bearded Couch G
Elytrigia atherica R Sea Couch-grass C
Elytrigia juncea ssp. boreoatlantica Sc Sand Couch C
Elytrigia repens C Common Couch G
* *Epilobium brunnescens* Sc New Zealand Willowherb U (Wa)
* *Epilobium ciliatum* R American Willowherb G
Epilobium ciliatum x parviflorum R G
Epilobium hirsutum C Great Willowherb W
Epilobium lanceolatum R Spear-leaved Willowherb G
Epilobium montanum C Broad-leaved Willowherb G
Eiplobium montanum x E. ciliatum R G
Eiplobium obscurum FC Short-fruited Willowherb G
Epilobium palustre FC Marsh Willowherb W
Epilobium parviflorum FC Hoary Willowherb G
0 *Epilobium roseum* R Pale Willowherb G

   Epilobium tetragonum FC Square-stalked Willowherb G
0 *Epipactis palustris* R Marsh Helleborine P (C)
   Equisetum arvense C Field Horsetail G
   Equisetum fluviatile Sc Water Horsetail W
   Equisetum palustre C Marsh Horsetail W
* *Erica ciliaris* R Dorset Heath M W
   Erica cinerea C Bell Heather M C
   Erica tetralix C Cross-leaved Heath M W
?* *Erica vagans* R Cornish Heath M C
0 *Erica x watsonii* = *E. ciliaris* x *E. tetralix* M R
* *Erigeron glaucus* Sc Seaside Daisy U Wa
* *Erigeron karvinskianus* Sc Mexican Fleabane U Wa
0* *Erigeron philadelphicus* Sc Robin's Plantain U
* *Erinus alpinus* R Fairy Foxglove U Wa
   Eriophorum angustifolium LC Common Cottongrass W
   Eriophorum vaginatum Sc Hare's-tail Cottongrass W
   Erodium cicutarium FC Common Stork's-bill C
   Erodium cicutarium ssp. dunense Sc C
   Erodium maritimum C Sea Stork's-bill C
   Erodium moschatum Sc Musk Stork's-bill C
   Erophila verna R Whitlow Grass G
0 *Eryngium campestre* R Field Eryngo F
   Eryngium maritimum Sc Sea-holly C
0* *Erysimum cheiranthoides* R Treacle Mustard F
0* *Erysimum cheiri* R Wallflower U
* *Escallonia macrantha* C Escallonia G
* *Escholzia californica* Sc Californian Poppy U
   Eupatorium cannabinum C Hemp-agrimony G(W)
0 *Euonymus europaeus* R Spindle H WD
   Euonymus japonicus R Evergreen Spindle G(C)
0 *Euphorbia exigua* R Dwarf Spurge F
   Euphorbia helioscopia C Sun Spurge F
   Euphorbia paralias Sc Sea Spurge C
   Euphorbia peplus C Petty Spurge G(F)
   Euphorbia portlandica Sc Portland Spurge C
0 *Euphrasia anglica* R Eyebright M
0 *Euphrasia confusa* R Eyebright M
   Euphrasia micrantha R Eyebright M
0 *Euphrasia nemorosa* R Eyebright M
   Euphrasia rostkoviana Sc Common Eyebright G

Appendix: PLANTS 159

   Euphrasia tetraquetra C Eyebright C
   Euphrasia vigursii R Vigur's Eyebright M
0 *Fagopyrum esculentum* R Buckwheat F
* *Fagus sylvatica* FC Beech WD H
* *Fallopia baldschuanica* Sc Russian-vine U
   Fallopia convolvulus C Black-bindweed G
* *Fallopia japonica* C Japanese Knotweed G
* *Fallopia sachalinensis* Sc Giant Knotweed G
* *Fascicularia pitcairniifolia* Sc C. Probably planted.
0 *Festuca arundinacea* C Tall Fescue G
   Festuca ovina C Sheep's-fescue C M
0 *Festuca pratensis* R Meadow Fescue G(P)
   Festuca rubra C Red Fescue M C P
   Festuca rubra ssp. juncea Sc C
* *Ficus carica* R Fig U
0 *Filago minima* Sc Small Cudweed G
0 *Filago vulgaris* R Common Cudweed G
   Filipendula ulmaria C Meadowsweet W
?* *Foeniculum vulgare* Sc Fennel G(H)
* *Forsythia x intermedia* Sc Forsythia G(U)
   Fragaria vesca Sc Wild Strawberry G
* *Fragaria x anassa* R Garden Strawberry G(U)
   Fraxinus excelsior C Ash WD H
* *Fuchsia magellanica* C Fuchsia G(U)
0* *Fuschia 'Riccartonii'* R G(U)
   Fumaria bastardii C Tall Ramping-fumitory F H
   Fumaria capreolata ssp. babingtonii LC White Ramping-fumitory G H
   Fumaria muralis ssp. boraei C Common Ramping-fumitory F H
   Fumaria occidentalis FC Western Ramping-fumitory F H
   Fumaria officinalis LC Common Fumitory F H
0 *Fumaria purpurea* R Purple Ramping-fumitory G
* *Galanthus nivalis* LC Snowdrop WD
0 *Galeopsis bifida* R Bifid Hemp-nettle G
   Galeopsis tetrahit FC Common Hemp-nettle G
* *Galinsoga quadriradiata* Sc Shaggy Soldier G(U)
   Galium aparine C Cleavers G
   Galium mollugo C Hedge Bedstraw H
   Galium odoratum R Woodruff G(U)
   Galium palustre C Common Marsh-bedstraw W
   Galium palustre ssp. elongatum Sc W

Galium palustre ssp. *palustre* C W
Galium saxatile C Heath Bedstraw M P
0 *Glaium spurium* R False Cleavers G. Record unconfirmed.
Galium verum C Lady's Bedstraw C H
0 *Gastridium ventricosum* C Lady's Bedstraw C H
0* *Gazania rigens* Sc Treasureflower U
* *Gaultheria mucronata* R Prickly Heath U
0 *Genista pilosa* R Hairy Greenweed C
Genista tinctoria ssp. *littoralis* SC Procumbent Dyer's Greenweed C
0 *Gentianella campestris* R Field Gentian P
0 *Geranium columbinum* Sc Long-stalked Crane's-bill G(P)
Geranium dissectum C Cut-leaved Crane's-bill G(P)
?* *Geranium lucidum* Sc Shining Crane's-bill G(H)
Geranium molle C Dove's-foot Crane's-bill G(P)
* *Geranium x oxonianum* = *G. endressii x G. versicolor* R Druce's Crane's-bill U
* *Geranium phaeum* R Dusky Crane's-bill G(U)
* *Geranium pratense* Sc Meadow Crane's-bill G(U)
Geraneum purpureum R Little-Robin G
0 *Geranium pusillum* FC Small-flowered Crane's-bill G(P)
* *Geranium pyrenaicum* Sc Hedgerow Crane's-bill G(U)
Geranium robertianum C Herb Robert G
Geranium rotundifolium Sc Round-leaved Crane's-bill G
* *Geranium sanguineum* Sc Bloody Crane's-bill U
* *Geranium versicolor* FC Pencilled Crane's-bill U(H)
Geum urbanum Sc Wood Avens WD
* *Gladiolus communis* ssp. *byzantinus* Sc Eastern Gladiolus G
0 *Glaucium flavum* R Yellow-horned Poppy C
0 *Glaux maritima* R Sea Milkwort C
Glechoma hederacea C Ground-ivy G(M)
Glyceria declinata FC Small Sweet-grass W
Glyceria fluitans FC Floating Sweet-grass W
Gnaphalium uliginosum FC Marsh Cudweed W F
* *Griselinia littoralis* Sc New Zealand Broadleaf U
* *Gunnera manicata* Sc Brazilian Giant-rhubarb W M
* *Gunnera tinctoria* C Giant-rhubarb W
0 *Gymnadenia borealis* R Fragrant Orchid M
* *Hebe x franciscana* = *H. elliptica x H. speciosa* Sc Hebe G(U)
Hedera helix ssp. *hibernica* C Atlantic Ivy G
0 *Helleborus viridis* ssp. *occidentalis* R Green Hellebore WD
* *Heracleum mantegazzianum* R Giant Hogweed G

Appendix: PLANTS

Heracleum sphondylium C Hogweed G
* *Hesperis matrionalis* Sc Dame's-violet G(U)
* *Hippophae rhamnoides* Sc Sea Buckthorn C
Hieracium umbellatum ssp. bichlorophyllum FC Leafy Hawkweed G(H)
Hieracium umbellatum ssp. umbellatum FC Narrow-leaved Hawkweed G(H)
* *Hoheria populnea* R New Zealand Mallow U
Holcus lanatus C Yorkshire-fog G
Holcus mollis FC Creeping Soft-grass G
Honckenya peploides LC Sea Sandwort C
Hordeum murinum Sc Wall Barley G
Humulus lupulus Sc Hop H C
* *Hyacinthoides hispanica* ?FC Spanish Bluebell U. Most records probably refer to a hybrid between this and the next species.
Hyacinthoides non-scripta C Bluebell C WD H
* *Hydrangea macrophylla* Sc Garden Hydrangea G(U)
Hydrocotyle vulgaris LC Marsh Pennywort W
0 *Hymenophyllum tunbridgense* R Tunbridge Filmy-fern M (Record is disputed by some botanists)
Hymenophyllum wilsonii R Wilson's Filmy-fern M
0 *Hyoscyamus niger* R Henbane C
Hypericum androsaemum FC Tutsan WD
* *Hypericum calcinum* R Rose-of-Sharon G
Hypericum elodes LC Marsh St John's-wort W
Hypericum humifusum FC Trailing St John's-wort M C
0 *Hypericum maculatum* R Imperforate St John's-wort G
Hypericum perforatum Sc Perforate St John's-wort G
Hypericum pulchrum FC Slender St John's-wort M C
Hypericum tetrapterum C Square-stalked St John's-wort G
Hypericum undulatum R Wavy St John's-wort W
Hypochaeris radicata C Cat's-ear G(H)
Ilex aquifolium C Holly M WD
Illecebrum verticillatum R Coral-necklace W
* *Impatiens glandulifera* LC Indian Balsam W
Inula crithmoides Sc Golden-samphire C
* *Inula helenium* R Elecampane G(U)
Iris foetididssima LC Stinking Iris C WD
Iris pseudacorus LC Yellow Iris W
Isolepis cernua Sc Slender Club-rush W
Isolepis setacea FC Bristle Club-rush W
Jasione montana C Sheep's-bit G

Juncus acutiflorus C Sharp-flowered Rush W
Juncus articulatus C Jointed Rush W
Juncus bufonius C Toad Rush W
Juncus bulbosus FC Bulbous Rush W
Juncus bulbosus ssp. kochii R M
0 *Juncus capitatus* R Dwarf Rush C M
Juncus conglomeratus C Compact Rush W
Juncus effusus C Soft-rush W
Juncus inflexus C Hard Rush W
Juncus maritimus R Sea Rush C
Juncus tenuis R Slender Rush G
Kickxia elatine Sc Sharp-leaved Fluellen F
Kickxia spuria R Round-leaved Fluellen F
Knautia arvensis FC Field Scabious F H
Koeleria macrantha FC Crested Hair-grass C
* *Lactuca saligna* R Least Lettuce C
* *Lactuca sativa* Sc Garden Lettuce G
* *Lagarosiphon major* R Curly Waterweed W
* *Lamiastrum galeobdolon ssp. argentatum* LC Variegated Yellow Archangel U
0 *Lamiastrum galeobdolon ssp. montanum* Yellow Archangel G
Lamium album Sc White Dead-nettle G
Lamium amplexicaule R Henbit Dead-nettle G
Lamium hybridum LC Cut-leaved Dead-nettle G
* *Lamium maculatum* C Spotted Dead-nettle G(U)
Lamium purpureum C Red Dead-nettle G
* *Lampranthus roseus* R Rosy Dew-plant U
0 *Lappula squarrosa* R Bur Foreget-me-not G
Lapsana communis C Nipplewort G(H)
0* *Lathyrus annuus* Sc Fodder Pea U
* *Lathyrus latifolius* R Broad-leaved Everlasting-pea U
Lathyrus linifolius var. montanus R Common Bitter-vetch WD H
Lathyrus pratensis FC Meadow Vetchling G
0 *Lathyrus sylvestris* R Narrow-leaved Everlasting Pea G
Laurus nobilis R Bay G
Lavatera arborea LC Tree-mallow C
Lavatera cretica R Smaller Tree-mallow C
Lemna minor C Common Duckweed W
* *Lemna minuta* FC Least Duckweed W
Lemna trisulca R Ivy-leaved Duckweed W
Leontodon autumnalis C Autumn Hawkbit G

Appendix: PLANTS 163

0 *Leontodon hispidus* R Rough Hawkbit P
    Leontodon saxatilis C Lesser Hawkbit G
0 *Lepidium campestre* R Field Pepperwort G
* *Lepidium draba* Sc Hoary Cress G
* *Lepidium sativum* R Garden Cress G
    Lepidium heterophyllum C Smith's Pepperwort G
0* *Leptospermum scoparium* R Broom Tea-tree
* *Leucanthemella serotina* Sc Autumn Oxeye G(U)
* *Leucanthemum x superbum* FC Shasta Daisy G(U)
    Leucanthemum vulgare C Oxeye Daisy C H
0* *Leucojum aestivum ssp. pulchellum* Sc Summer Snowflake U
    Ligustrum vulgare Sc Wild Privet G
* *Ligustrum ovalifolium* C Garden Privet G(U)
    Limonium loganicum R Rock Sea-lavender C
* *Linaria x dominii* = *L. purpurea x L. repens* R G(U)
* *Linaria purpurea* Sc Purple Toadflax U
    Linaria repens R Pale Toadflax G
0 *Linaria supina* R Prostrate Toadflax G
    Linaria vulgaris C Common Toadflax G(H)
    Linum bienne Sc Pale Flax F P
    Linum catharticum Sc Fairy Flax F P
* *Linum usitatissimum* FC Flax G
    Littorella uniflora Sc Shoreweed W
* *Lobularia maritima* Sc Sweet Alison U
* *Lolium multiflorum* C Italian Rye-grass F
    Lolium perenne C Perennial Rye-grass G
* *Lonicera japonica* Sc Japanese Honeysuckle G(H)
* *Lonicera nitida* R Wilson's Honeysuckle U
    Lonicera periclymenum C Honeysuckle G(H)
    Lotus angustissimus R Slender Bird's-foot-trefoil C
    Lotus corniculatus C Common Bird's-foot-trefoil C H
    Lotus pedunculatus C Greater Bird's-foot-trefoil W
    Lotus subbiflorus LC Hairy Bird's-foot-trefoil C
* *Lunaria annua* FC Honesty U
* *Luma apiculata* R Chilean Myrtle U
    Luzula campestris C Field Wood-rush C M
    Luzula multiflora FC Heath Wood-rush M
    Luzula sylvatica Sc Great Wood-rush M WD
    Lychnis flos-cuculi LC Ragged-Robin W
* *Lycopersicon esculentum* Sc Tomato G(U)

Common Toadflax
Linaria vulgaris

    Lycopus europaeus Sc Gypsywort W
*   *Lysichiton americanus* Sc American Skunk-cabbage W
    Lysimachia nemorum Sc Yellow Pimpernel WD
0* *Lysimachia nummularia* R Creeping Jenny U
*   *Lysimachia punctata* R Dotted Loosestrife U
    Lysimachia vulgaris R Yellow Loosestrife W
    Lythrum portula C Water-purslane W
    Lythrum portula ssp. longidentata R WD
    Lythrum salicaria C Purple-loosestrife W
*   *Malus domestica* Sc Apple G
*   *Malus sylvestris* R Crab Apple G(U)
    Malva moschata R Musk-mallow G
    Malva neglecta R Dwarf Mallow C
*   *Malva nicaeensis* R French Mallow G
    Malva sylvestris C Common Mallow G
*   *Matricaria discoidea* FC Pineapple-weed F
    Matricaria recutita R Scented Mayweed G
0* *Meconopsis cambrica* R Welsh Poppy U
    Medicago arabica C Spotted Medick G(P)
    Medicago lupulina C Black Medick G(P)
    Medicago polymorpha LC Toothed Medick G(P)
0* *Medicago sativa ssp. falcata* R Sickle Medick G
*   *Medicago sativa ssp. varia* R Sand Lucerne Wa
0   *Melampyrum pratense ssp. pratense* R Common Cow-wheat WD H
0* *Melilotus albus* R White Melilot G(P)
0* *Melilotus altissimus* R Tall Melilot G(P)
0* *Melilotus indicus* R Small Melilot G(P)
*   *Melilotus officinalis* R Ribbed Melilot G
0* *Melissa officinalis* LC Balm G
    Mentha aquatica C Water Mint W
    Mentha arvensis Sc Corn Mint F
0   *Mentha x gracilis = M. arvensis x M. spicata* R G
    Mentha x piperita = M. aquatica x M. spicata R Peppermint G
0   *Mentha pulegium* R Pennyroyal W
0   *Mentha x smithiana = M. arvensis x M. aquatica x M. spicata* R Tall Mint G
*   *Mentha spicata* LC Spear Mint G
    Mentha suaveolens FC Round-leaved Mint G
0   *Mentha x suavis = M. aquatica x M. suaveolens* FC G
    Mentha x verticillata = M. aquatica x M. arvensis Sc Whorled Mint G
    Mentha x villosa = M. spicata x M. suaveolens R Apple Mint G

Appendix: PLANTS

    Menyanthes trifoliata Sc Bogbean W
    Mercurialis perennis FC Dog's Mercury WD
0* *Mespilus germanica* R Medlar G
0* *Mimulus guttatus* Sc Monkeyflower W
    Mimulus x robertsii = M. guttatus x M. luteus Sc Hybrid Monkeyflower W
    Misopates orontium FC Weasel's-snout G
    Moehringia trinervia Sc Three-nerved Sandwort WD H
    Moenchia erecta Sc Upright Chickweed C
    Molinia caerulea C Purple Moor-grass M
    Montia fontana C Blinks W
    Montia fontana ssp. amporitana R F
* *Morus nigra* R Black Mulberry G. Possibly planted.
    Myosotis arvensis C Field Forget-me-not G
    Myosotsis dicolor C Changing Forget-me-not G
    Myosotsis laxa ssp. cespitosa Sc Tufted Forget-me-not W
    Myosotsis ramosissima Sc Early Forget-me-not G
    Myosotsis secunda FC Creeping Forget-me-not W
    Myosotes sylvatica FC Wood Forget-me-not G
0 *Myrica gale* R Bog Myrtle W
    Myriophyllum alterniflorum Sc Alternate Water-milfoil W
* *Myriophyllum aquaticum* Sc Parrot's-feather U(W)
* *Mrytus communis* R Common Myrtle U
* *Narcissus cultivars* C Daffodils WD C H
0 *Narcissus x medioluteus* R Primrose-peerless G
* *Narcissus pseudonarcissus* Sc Wild Daffodil C
0 *Nardus stricta* R Mat-grass M
    Narthecium ossifragum LC Bog Asphodel M W
* *Nicandra physaloides* R Apple-of-Peru G(U)
* *Nigella damascena* Sc Love-in-a-mist G(U)
* *Nuphar lutea* R Yellow Water-lily W
* *Nymphaea alba* Sc White Water-lily W
* *Nymphoides peltata* R Fringed Water-lily W
0 *Odontites vernus ssp. serotinus* R Red Bartsia G
    Oenanthe crocata C Hemlock Water-dropwort W
0* *Oenothera biennis* R Common Evening-primrose G(U)
0* *Oenothera cambrica* R Small-flowered Evening-primrose G(U)
* *Oenothera glazioviana* Sc Large-flowered Evening-primrose G(U)
0* *Oenothera stricta* R Fragrant Evening-primrose U
* *Olearia macrodonta* R New Zealand Holly H
* *Olearia solandri* R Coastal Daisybush U

* *Olearia traversii* Sc Ake ake G(U)
  Ononis repens FC Common Restharrow G
* *Onopordon acanthium* Sc Cotton Thistle G
0 *Ophrys apifera* R Bee Orchid P
  Orchis mascula FC Early-purple Orchid WD H
0 *Origanum vulgare* R Wild Marjoram G
* *Ornithogalum angustifolium* R Star-of-Bethlehem U WD
  Ornithopus perpusillus C Bird's-foot G(P)(C)
  Orobanche minor Sc Common Broomrape G
0 *Orobanche rapum-genistae* R Greater Broomrape G(M)
  Osmunda regalis LC Royal Fern W
  Oxalis acetosella C Wood Sorrel M WD
* *Oxalis articulata* FC Pink-sorrel U
* *Oxalis corniculata* FC Procumbent Yellow-sorrel G
* *Oxalis incarnata* FC Pale Pink-sorrel U
* *Oxalis latifolia* FC Garden Pink-sorrel U
* *Oxalis megalorrhiza* R Fleshy Yellow-sorrel U
* *Oxalis pes-caprae* R Bermuda-buttercup U H
* *Oxalis rosea* R Annual Pink-sorrel U
0* *Oxalis stricta* Sc Upright Yellow-sorrel U
0?**Papaver argemone* R Prickly Poppy C
?* *Papaver dubium* Sc Long-headed Poppy G
0?**Papaver hybridum* R Rough Poppy G
  Papaver rhoeas FC Common Poppy F
* *Papaver somniferum* Sc Opium Poppy G
* *Paspalum distichum* R Water Finger-grass U(W)
  Parentucellia viscosa Sc Yellow Bartsia W F
  Parietaria judaica C Pellitory-of-the-wall Wa
0 *Pastinaca sativa var. sylvestris* R Wild Parsnip G(P)
  Pedicularis palustris Sc Marsh Lousewort W
  Pedicularis sylvatica ssp. sylvatica C Lousewort M C
* *Pentaglottis sempervirens* FC Green Alkanet G(U)
* *Pericallis hybrida* LC Cineraria U
  Persicaria amplexicaulis Sc Red Bistort U
* *Persicaria campanulata* Sc Lesser Knotweed G
  Persicaria hydropiper C Water-pepper W
  Persicaria lapathifolia FC Pale Persicaria F
  Persicaria maculosa C Redshank W
* *Persicaria wallichii* Sc Himalayan Knotweed G
* *Petasites fragrans* C Winter Heliotrope U

Appendix: PLANTS

- * *Petasites hybridus* Sc Butterbur G(U)
- * *Petroselinum crispum* Sc Garden Parsley U
- * *Phacelia tanacetifolia* R Phacelia F
- 0 *Phalaris arundinacea* R Reed Canary-grass W
- * *Phalaris canariensis* Sc Canary-grass G(U)
- *Phleum bertolonii* FC Smaller Cat's-tail G(C)
- *Phleum pratense* C Timothy G
- *Phragmites australis* C Common Reed W
- *Phyllitis scolopendrium* C Hart's-tongue Fern WD Wa
- * *Phytolacca acinosa* R Indian Pokeweed U
- * *Picea sitchensis* Sc Sitka spruce WD
- ?* *Picris echioides* FC Bristly Oxtongue G
- * *Pilosella aurantiacum ssp. carpathicola* R Fox-and-cubs G(U)
- *Pilosella officinarum* FC Mouse-ear-hawkweed G
- *Pilosella officinarum ssp. melanops* R G
- *Pilularia globulifera* R Pillwort W
- *Pimpinella saxifraga* R Burnet-saxifrage G(H)
- *Pinguicula lusitanica* Sc Pale Butterwort W
- *Pinus pinaster* R Maritime Pine G
- * *Pinus radiata* FC Monterey Pine G
- * *Pinus sylvestris* Sc Scots Pine WD
- * *Pittosporum crassifolium* FC Karo G(U)
- * *Pittosporum ralphii* R U
- * *Pittosporum tenuifolium* Sc Kohuhu G(U)
- *Plantago coronopus* C Buck's-horn Plantain C
- *Plantago lanceolata* C Ribwort Plantain G
- *Plantago major* C Greater Plantain G
- *Plantago maritima* C Sea Plantain C
- *Platanthera bifolia* R Lesser Butterfly-orchid M
- *Poa annua* C Annual Meadow-grass G
- 0 *Poa humulis* Sc Spreading Meadow-grass C G(F)
- *Poa infirma* R Early Meadow-grass C
- 0 *Poa nemoralis* R Wood Meadow-grass Wa H
- *Poa pratensis* C Smooth Meadow-grass G(F)
- *Poa trivialis* C Rough Meadow-grass G(F)
- *Polygala serpyllifolia* FC Heath Milkwort MC
- *Polygala vulgaris* C Common Milkwort M C
- * *Polygonatum x hybridum* R Garden Solomon's-seal WD H
- *Polygonum arenastrum* C Equal-leaved Knotgrass F
- *Polygonum aviculare* C Knotgrass G

Ribwort Plantain
Plantago lanceolata

0   *Polygonum oxyspermum ssp. raii* R Ray's Knotgrass C
    Polypodium interjectum C Intermediate Polypody WD H
    Polypodium vulgare FC Polypody WD H Wa
*   *Polypogon viridis* R Water Bent G
    Polystichum setiferum Sc Soft Shield-fern WD
*   *Pontederia cordata* R Pickerelweed W
0* *Populus alba* R White Poplar U
*   *Populus x canadensis = P. deltoides x P. nigra* Sc Hybrid Black-poplar G(U)
*   *Populus tremula* R Aspen U
    Potamogeton berchtoldii R Small Pondweed W
0   *Potamogeton natans* R Broad-leaved Pondweed W
0   *Potamogeton pectinatus* R Fennel Pondweed W
    Potamogeton polygonifolius C Bog Pondweed W
0   *Potamogeton pusillus* R Lesser Pondweed W
    Potentilla anglica C Trailing Tormentil G(H)
    Potentilla anserina C Silverweed G
    Potentilla erecta C Tormentil M C
0   *Potentilla palustris* Sc Marsh Cinquefoil W
*   *Potentilla recta* R Sulphur Cinquefoil G
    Potentilla reptans C Creeping Cinquefoil G(H)
    Potentilla sterilis C Barren Strawberry G
0   *Potentilla x suberecta = P. erecta x P. anglica* Sc G(F)
*   *Primula veris* R Cowslip G(U)
    Primula vulgaris C Primrose C WD H
    Prunella vulgaris C Selfheal G
0* *Prunus avium* R Wild Cherry G
*   *Prunus cerasus* R Dwarf Cherry H WD
*   *Prunus domestica* R Wild Plum H
0* *Prunus domestica ssp. institia* R Bullace H
0* *Prunus dulcis* R Almond U
*   *Prunus laurocerasus* R Cherry-laurel U WD
*   *Prunus serratula* R Japanese Cherry U
    Prunus spinosa C Blackthorn M C H
*   *Pseudosasa japonica* Sc Arrow-bamboo G(U)
    Pteridium aquilinum C Bracken M C
0* *Pulmonaria officinalis* Sc Lungwort G(U)
    Pulicaria dysenterica C Common Fleabane G(W)
*   *Quercus cerris* FC Turkey Oak WD
*   *Quercus ilex* FC Evergreen Oak WD
    Quercus petraea C Sessile Oak WD

Appendix: PLANTS 169

Quercus robur C Pedunculate Oak WD
Quercus x rosacea = *Q. petraea x Q. robur* R Oak hybrid WD
Radiola linoides LC Allseed G
Ranunculus acris C Meadow Buttercup G(F)
0 *Ranunculus bulbosus* R Bulbous Buttercup F
Ranunculus ficaria C Lesser Celandine G (WD)(H)
Ranunculus ficaria ssp. bulbifer R Lesser Celandine G(WD)(H)
Ranunculus flammula ssp. flammula C Lesser Spearwort W
Ranunculus hederaceus C Ivy-leaved Crowfoot W
Ranunculus omiophyllus FC Round-leaved Crowfoot W
Ranunculus parviflorus Sc Small-flowered Buttercup F
Ranunculus repens C Creeping Buttercup G
Ranunculus tripartitus R Three-lobed Crowfoot W
Raphanus raphanistrum ssp. maritimus FC Sea Radish C
Raphanus raphanistrum ssp. raphanistrum C Wild Radish G
Reseda lutea R Wild Mignonette G
Reseda luteola FC Weld G
Rhinanthus minor FC Yellow-rattle G(P)
* *Rhododendron ponticum* C Rhododendron WD M U
0 *Rhynchospora alba* R White Beak-sedge W
?* *Ribes rubrum* Sc Red Currant G
* *Ribes nigrum* R Black Currant G(U)
?* *Ribes uva-crispa* Sc Gooseberry G
Rorippa nasturtium-aquaticum FC Water-cress W
Rorippa palustris R Marsh Yellow-cress W
0 *Rorippa x sterilis* = *R. nasturtium-aquaticum x R. microphylla* R W
Rosa arvensis FC Field-rose H
Rosa canina FC Dog-rose H
Rosa x dumalis = *R. canina x R. caesia* R G
Rosa pimpinellifolia Sc Burnet Rose C
* *Rosa rugosa* Sc Japanese Rose U
0 *Rosa stylosa* R Short-styled Field-rose H
Rubus agg. C Bramble G
Rubia peregrina FC Wild Madder G(H)
Rumex acetosa C Common Sorrel G(H)
Rumex acetosa ssp. biformis R Common Sorrel C
Rumex acetosella C Sheep's Sorrel M
Rumex conglomeratus C Clustered Dock G(F)
Rumex crispus C Curled Dock G(F)
0 *Rumex hydrolapathum* R. Water Dock C

Rumex obtusifolius C Broad-leaved Dock G(F)
Rumex pulcher ssp. pulcher C Fiddle Dock G(F)
Rumex rupestris R Shore Dock C
Rumex sanguineus C Wood Dock WD
0 *Ruppia maritima* R Beaked Tasselweed W
0 *Ruscus aculeatus* Sc Butcher's-broom WD
Sagina apetala C Annual Pearlwort G
Sagina apetala ssp. apetala R Ciliate Pearlwort G
Sagina apetala ssp. erecta C Annual Pearlwort G(P)
Sagina maritima C Sea Pearlwort C
Sagina nodosa R Knotted Pearlwort G
Sagina procumbens C Procumbent Pearlwort G(P)
Sagina subulata C Heath Pearlwort M
* *Sagittaria sagittifolia* R Arrowhead W
?* *Salix alba* Sc White Willow W
Salix x ambigua = S. aurita x S. repens Sc M
Salix aurita C Eared Willow M W
0* *Salix caprea* R Goat Willow H
Salix cinerea ssp. oleifolia C Grey Willow M W
0* *Salix fragilis* R Crack Willow U
Salix repens Sc Creeping Willow M
* *Salix x sepulcralis* Sc Weeping Willow G. Possibly planted.
Salix x sericans = S. caprea x S. viminalis Sc Broad-leaved Osier U
Salix viminalis Sc Osier W
0 *Salsola kali ssp. kali* R Prickly Saltwort C
Sambucus nigra C Common Elder G
Samolus valerandi FC Brookweed C
Sanguisorba minor ssp. minor R Salad Burnet C
Sanicula europaea R Sanicle WD
* *Saponaria officinalis* Sc Soapwort G(U)
* *Sasa palmata* Sc Broad-leaved Bamboo G(U)
* *Saxifraga cymbalaria* R Celandine Saxifrage U
* *Saxifraga x urbium = S. umbrosa x S. spathularis* Sc Londonpride G(U)
0 *Scandix pecten veneris* R Shepherd's-needle H F
Schoenoplectus tabernaemontani Sc Grey Club-rush W
Schoenus nigricans LC Black Bog-rush W
0 *Scilla autumnalis* Sc Autumn Squill C
Scilla verna C Spring Squill C
0 *Scleranthus annuus ssp. annuus* R Annual Knawel M F
Scrophularia auriculata C Water Figwort W

Appendix: PLANTS

    Scrophularia nodosa C Common Figwort G
    Scrophularia scorodonia Sc Balm-leaved Figwort G(W)
    Scutellaria galericulata Sc Skull-cap W
    Scutellaria minor Sc Lesser Skull-cap W
0* *Securigera varia* R Crown Vetch G
    Sedum acre FC Biting Stonecrop G(U)(Wa)
* *Sedum album* Sc White Stonecrop U
    Sedum anglicum C English Stonecrop C H
0* *Sedum confusum* Sc Lesser Mexican-stonecrop U
* *Sedum forsteranum* R Rock Stonecrop G
* *Sedum rupestre* Sc Reflexed Stonecrop U Wa
* *Sedum spurium* R Caucasian Stonecrop G(U)
    Sedum telephium R Orpine C
* *Selaginella kraussiana* R Krauss's Clubmoss U(WD)(H)
    Senecio aquaticus FC Marsh Ragwort W
* *Senecio cineraria* FC Silver Ragwort G(U)
    Senecio jacobaea C Common Ragwort G
    Senecio x ostenfeldii = *S. jacobaea x S. aquaticus* LC G(P)
* *Senecio squalidus* Sc Oxford Ragwort G(U)
    Senecio sylvaticus FC Heath Groundsel G(M)
* *Senecio vernalis* R Eastern Groundsel U
    Senecio vulgaris C Groundsel G
    Serratula tinctoria C Saw-wort C M
0* *Setaria viridis* Sc Green Bristle-grass Sc
    Sherardia arvensis FC Field Madder H
    Sibthorpia europaea LC Cornish Moneywort WD W
    Silene dioica C Red Campion G
0 *Silene gallica var. anglica* R Small-flowered Catchfly F
    Silene x hampanea = *S. latifolia x S. dioica* R G
0 *Silene latifolia* Sc White Campion G
    Silene noctiflora R Night-flowering Catchfly F
    Silene uniflora C Sea Campion C
    Silene vulgaris ssp. vulgaris R Bladder Campion G
* *Silybum marianum* R Milk Thistle G
0* *Sinapis alba* R White Mustard G
    Sinapis arvensis C Charlock F
* *Sisymbrium altissimum* Sc Tall Rocket G
    Sisymbrium officinale C Hedge Mustard H
* *Sisymbrium orientale* R Eastern Rocket G
* *Smyrnium olusatrum* C Alexanders G(H)

    Solanum dulcamara C Bittersweet G(H)
* *Solanum laciniatum* Sc Kangaroo-apple U
    Solanum nigrum FC Black Nightshade C F
* *Solanum tuberosum* FC Potato G(F)
* *Soleirolia soleirolii* C Mind-your-own-business U
* *Solidago canadensis* Sc Canadian Goldenrod G(U)
    Solidago virgaurea FC Goldenrod C M H
    Sonchus arvensis FC Perennial Sow-thistle G
    Sonchus asper C Prickly Sow-thistle G
    Sonchus oleraceus C Smooth Sow-thistle G
* *Sorbus aucuparia* R Rowan G(U)
* *Sorbus intermedia agg.* R Swedish Whitebeam U
    Sparganium emersum R Unbranched Bur-reed W
    Sparganium erectum FC Branched Bur-reed W
    Spergula arvensis C Corn Spurrey F
* *Spergularia bocconii* R Greek Sea-spurrey C. Disappeared 1989.
0 *Spergularia marina* R Lesser Sea-spurrey C
0 *Spergularia media* R Greater Sea-spurrey C
    Spergularia rubra FC Sand Spurrey G
    Spergularia rupicola C Rock Sea-spurrey C
0* *Spiraea salicifolia* R Bridewort G
    Spiranthes spirales LC Autumn Lady's-tresses P(C)
    Stachys x ambigua = S. sylvatica x S. palustris Sc Hybrid Woundwort G
    Stachys arvensis FC Field Woundwort W
    Stachys officinalis C Betony M C H
    Stachys palustris FC Marsh Woundwort W
    Stachys sylvatica C Hedge Woundwort G
    Stellaria graminea C Lesser Stitchwort H
    Stellaria holostea C Greater Stitchwort G(H)
    Stellaria media C Common Chickweed G(F)
    Stellaria neglecta R Greater Chickweed H
    Stellaria pallida R Lesser Chickweed C
    Stellaria uliginosa C Bog Stitchwort W
    Succisa pratensis LC Devil's-bit Scabious M W
0* *Symphoricarpos albus* Sc Snowberry G(U)
    Symphytum officinale FC Common Comfrey G(U)
?* *Symphytum tuberosum* R Tuberous Comfrey G
* *Symphytum x uplandicum* Sc Russian Comfrey G
* *Tagetes patula* R French Marigold U
    Tamus communis FC Black Bryony WD H

Appendix: PLANTS

* *Tamarix gallica* FC Tamarisk C
* *Tanacetum parthenium* FC Feverfew G(U)
* *Tanacetum vulgare* FC Tansy G(U)
   Taraxacum officinale agg. C Dandelion G
0* *Taxus baccata* Sc Yew G
0   *Teesdalia nudicaulis* R Shepherd's Cress M H Wa
* *Tellima grandiflora* R Fringecups U
* *Teucrium chamaedrys* R Wall Germander U(Wa)
   Teucrium scorodonia C Wood Sage M
   Thalictrum minus R Lesser Meadow-rue M
   Thelypteris palustris R Marsh Fern W. Believed extinct.
?* *Thlaspi arvense* LC Field Penny-cress F
   Thymus polytrichus C Wild Thyme C M
* *Tilia x europeaea* R Lime WD U
* *Tolmiea menziesii* R Pick-a-back-plant U
   Torilis japonica Sc Upright Hedge-parsley H
   Torilis nodosa Sc Knotted Hedge-parsley C
0* *Tragopogon porrifolius* R Salsify C
0   *Tragopogon pratensis ssp. minor* R Goat's-beard G
   Trichophorum cespitosum ssp. germanicum R Deergrass M
   Trifolium arvense FC Hare's-foot Clover G
   Trifolium campestre C Hop Trefoil G(P)
   Trifolium dubium C Lesser Trefoil G(P)
0   *Trifolium fragiferum* R Strawberry Clover C
* *Trifolium hybridum* LC Alsike Clover G
* *Trifolium incarnatum ssp. incarnatum* Sc Crimson Clover G(F)
   Trifolium medium FC Zigzag Clover G(P)
   Trifolium micranthum C Slender Trefoil G(P)
   Trifolium occidentale FC Western Clover C
   Trifolium ornithopodioides Sc Bird's-foot Clover C
   Trifolium pratense C Red Clover G(P)
   Trifolium repens C White Clover G(P)
   Trifolium scabrum FC Rough Clover G(P)(C)
0   *Trifolium squamosum* R Sea Clover C
   Trifolium striatum Sc Knotted Clover G(P)
   Trifolium subterraneum C Subterranean Clover G(P)
   Triglochin maritima R Sea Arrowgrass C
   Triglochin palustre R Marsh Arrowgrass W
   Tripleurospermum inodorum C Scentless Mayweed F
   Tripleurospermum maritimum C Sea Mayweed C

0   *Trisetum flavescens* R Yellow Oat-grass G(P)
*   *Tropaeolum majus* R Nasturtium G U
*   *Tsuga heterophylla* Sc Western Hemlock-spruce WD
    Tussilago farfara R Colt's-foot G
    Typha latifolia Sc Bulrush W
    Ulex europaeus C Gorse C M
    Ulex gallii C Western Gorse C M
    Ulmus glabra C Wych Elm WD
?* *Ulmus minor ssp. angustifolia* FC Cornish Elm WD H
    Umbilicus rupestris C Navelwort U H Wa
    Urtica dioica C Common Nettle G
    Urtica urens FC Small Nettle F
0   *Utricularia australis* R Bladderwort W
0   *Utricularia minor* R Lesser Bladderwort W.
    Vaccinium myrtillus C Bilberry M
    Valerianella carinata Sc Keel-fruited Cornsalad G
0   *Valerianella dentata* R Narrow-fruited Cornsalad G
0   *Valerianella eriocarpa* R Hairy-fruited Cornsalad G
0   *Valerianella locusta* R Common Cornsalad G
0   *Valerianella rimosa* R Broad-fruited Cornsalad G
0   *Verbascum nigrum* R Dark Mullein C(G)
    Verbascum thapsus FC Great Mullein G
0   *Verbascum virgatum* R Twiggy Mullein G
    Verbena officinalis FC Vervain G
    Veronica agrestis R Green Field-speedwell G(F)
    Veronica arvensis FC Wall Speedwell G(Wa)
    Veronica beccabunga FC Brooklime W
    Veronica chamaedrys C Germander Speedwell G
*   *Veronica filiformis* FC Slender Speedwell G
    Veronica hederifolia ssp. hederifolia C Ivy-leaved Speedwell G
    Veronica hederifolia ssp. lucorum C Ivy-leaved Speedwell G
    Veronica montana C Wood Speedwell WD
    Veronica officinalis FC Heath Speedwell G
*   *Veronica peregrina* LC American Speedwell G(U)
*   *Veronica persica* C Common Field-speedwell G(F)
    Veronica polita R Grey Field-speedwell G(F)
    Veronica scutellata LC Marsh Speedwell W
    Veronica serpyllifolia C Thyme-leaved Speedwell G
0* *Vicia bithynica* R Bithynian Vetch H
    Vicia cracca C Tufted Vetch G(H)

    Vicia hirsuta C Hairy Tare G
0* *Vicia pannonica* Sc Hungarian Vetch U
    Vicia sativa ssp. nigra C Common Vetch G(P)
* *Vicia sativa ssp. sativa* FC Common Vetch G(P)
    Vicia sepium FC Bush Vetch G(H)
0 *Vicia sylvatica* R Wood Vetch G
    Vicia tetrasperma Sc Smooth Tare G
0* *Vicia villosa* Sc Fodder Vetch G
* *Vinca major* C Greater Periwinkle U H
* *Vinca minor* FC Lesser Periwinkle U H
    Viola arvensis FC Field Pansy F
0 *Viola canina ssp. canina* R Heath dog-violet C M
* *Viola cornuta* R Horned Pansy U
    Viola lactea Sc Pale Dog-violet C M
* *Viola odorata* FC Sweet Violet G
    Viola palustris ssp. juressi Sc Marsh Violet W
    Viola riviniana C Common Dog-violet G(C)(M)
    Viola riviniana x Viola lactea Sc M C
    Viscum album R Mistletoe U
    Vulpia bromoides FC Squirrel-tail Fescue C
    Vulpia myuros Sc Rat's-tail Fescue G
* *Yucca gloriosa* R Spanish Dagger U
* *Yushania anceps* R Indian Fountain-bamboo U
    Wahlenbergia hederacea Sc Ivy-leaved Bellflower W
* *Zantedeschia aethiopica* R Altar-lily W
* *Zelkova serrata* R Japanese Zelkova U
    Zostera marina Sc Eelgrass W

Appendix

BIRDS OF THE LAND'S END PENINSULA

The following is a list of all the species which have been recorded on the peninsula, including records from Mounts Bay which extends westwards to Mousehole, and also St Ives Bay. The dates of very old records are given. There have been many escapes of exotic birds over the years in and around the district, but few of these are included in the list, one example being a species of Sunbird which frequented gardens at Porth Curno and Porthgwarra in the summer of 1995.

The principal sources of reference were *Birds of the Cornish Coast* and *Birds of Cornwall* by R. H. Penhallurick, and the Cornwall Birdwatching and Preservation Society reports 1976-1999. The list is up to date, but excludes a few species pending acceptance or rejection by the Rarities Committee. Further details of localities and recorders may be obtained from the references above.

The following codes refer to status in the peninsula:
B = Breeds   Re = Resident   Vs = Visitor
PM = Passage Migrant   Vg = Vagrant   S = Summer
W = Winter   U = Uncommon   Sc = Scarce
R = Rare   V = Very rare

Names in inverted commas refer to races rather than a separate species.

Red-throated Diver *Gavia stellata* PM.WVs
Black-throated Diver *Gavia arctica* PM.WVs
Great Northern Diver *Gavia immer* PM.WVs
White-billed Diver *Gavia adamsii* VVg
Little Grebe *Tachybaptus ruficollis* B?.Re.WVs
Great Crested Grebe *Podiceps cristatus* ScPM and WVs
Red-necked Grebe *Podiceps grisegena* ScPM and WVs
Slavonian Grebe *Podiceps auritus* ScPM and WVs
Black-necked Grebe *Podiceps nigricollis* ScPM and WVs

Appendix: BIRDS 177

Black-browed Albatross *Diomedia melanophris* VVg
Northern Fulmar *Fulmaris glacialis* B.Re.PM
Fea's Petrel *Pterodroma feae* VVg
Cory's Shearwater *Calonectris diomedea* ScPM
Great Shearwater *Puffinus gravis* ScPM
Sooty Shearwater *Puffinus griseus* PM
Manx Shearwater *Puffinus puffinus* PM. Feeding parties in summer.
Balearic Shearwater *Puffinus yelkouan* UPM
Little Shearwater *Puffinus assimilis* VVg
Wilson's Storm-Petrel *Oceanites oceanicus* VVg
European Storm Petrel *Hydrobates pelagicus* UPM. Small feeding parties. Probably has bred.
Leach's Storm-Petrel *Oceanodroma leucorhoa* ScPM
Northern Gannet *Morus bassanus* PM. Feeding parties.
Great Cormorant *Phalacrocorax carbo* SVs.WVs
Shag *Phalacrocorax aristotelis* B.R.PM
Great Bittern *Botaurus stellaris* RWVs
American Bittern *Botaurus lentiginosus* VVg (1906)
Little Bittern *Ixobrychus minutus* RVg
Night Heron *Nycticorax nycticorax* RVg
Squacco Heron *Ardeola ralloides* VVg
Cattle Egret *Bulbulcus ibis* VVg
Little Egret *Egretta garzetta* PM.WVs
Great Egret *Egretta alba* VVg (1999)
Grey Heron *Ardea cinerea* PM.WVs.SVs
Purple Heron *Ardea purpurea* Vg
Black Stork *Ciconia nigra* VVg
White Stork *Ciconia ciconia* VVg
Glossy Ibis *Plegadis falcinellus* VVg
Eurasian Spoonbill *Platalea leucorodia* RPM
Flamingo *Phoenicopterus sps.* VVg (1912)
Mute Swan *Cygnus olor* B SVs.Wvs
Bewick's Swan (Tundra Swan) *Cygnus columbianus* ScPM.WVs
Whooper Swan *Cygnus cygnus* ScPM and WVs
Bean Goose *Anser fabalis* RPM
Pink-footed Goose *Anser brachyrhynchus* ScPM
White-fronted Goose *Anser albifrons* ScPM
Grey Lag Goose *Anser anser* ScPM
Snow Goose *Anser caerulescens* VVg
Canada Goose *Branta canadensis* ScPM

Barnacle Goose *Branta leucopsis* UPM
Brent Goose *Branta bernicla* UPM
Egyptian Goose *Alopochen aegyptiacus* Feral escape.
Ruddy Shelduck *Tadorna ferruginea* Vg or escape.
Common Shelduck *Tadorna tadorna* PM
American Wood Duck *Aix sponsa* VVg or more likely an escape.
Eurasian Wigeon *Anas penelope* PM. WVs
American Wigeon *Anas americana* RVg
Gadwall *Anas strepera* ScPM and WVs
Common Teal *Anas crecca* PM.WVs
Green-winged Teal *Anas carolinensis* RVg
Mallard *Anas platyrhynchos* B.Re.WVs
Northern Pintail *Anas acuta* ScPM and WVs
Garganey *Anas querquedula* ScPM
Blue-winged Teal *Anas discors* RVg
Northern Shoveler *Anas clypeata* ScPM and WVs
Common Pochard *Aythya ferina* PM.WVs
Ring-necked Duck *Aythya collaris* RVg
Ferruginous Duck *Aythya nyroca* VVg
Tufted Duck *Aythya fuligula* PM.WVs
Greater Scaup *Aythya marila* UPM and WVs
Lesser Scaup *Aythya affinis* RVg
Common Eider *Somateria mollissima* UPM and WVs
King Eider *Somateria spectabilis* RVs (Mounts Bay only)
Long-tailed Duck *Clangula hyemalis* UPM and WVs
Common Scoter *Melanitta nigra* PM
Surf Scoter *Melanitta perspicillata* VVg
Velvet Scoter *Melanitta fusca* UPM
Common Goldeneye *Bucephala clangula* ScPM and WVs
Smew *Mergellus albellus* RPM and WVs
Red-breasted Merganser *Mergus serrator* UPM
Goosander *Mergus merganser* UPM
Ruddy Duck *Oxyura jamaicensis* RWVs
Honey Buzzard *Pernis apivorus* RPM
Black Kite *Milvus migrans* RPM
Red Kite *Milvus milvus* RPM
White-tailed Eagle *Haliaeetus albicilla* VVg. 1948
Marsh Harrier *Circus aeruginosus* RPM
Hen Harrier *Circus cyaneus* ScPM and WVs
Montagu's Harrier *Circus pygargus* RPM

Appendix: BIRDS 179

Northern Goshawk *Accipiter gentilis* RPM
Eurasian Sparrow Hawk *Accipiter nisus* B.Re.PM
Common Buzzard *Buteo buteo* B.RE.PM
Rough-legged Buzzard *Buteo lagopus* RPM
Golden Eagle *Aquila chrysaetos* VVg. 1859
Booted Eagle *Hieraaetus pennatus* VVG 1999. Record pending acceptance by Rarities Committee.
Osprey *Pandion haliaetus* RPM
Lesser Kestrel *Falco naumanni* VVg
Common Kestrel *Falco tinnunculus* B.Re.PM
Red-footed Falcon *Falco vespertinus* RVg
Merlin *Falco columbarius* ScPM and WVs
Hobby *Falco subbuteo* ScPM
Saker Falcon *Falco cherrug* VVg or escape
Gyr Falcon *Falco rusticolus* VVg
Peregrine Falcon *Falco peregrinus* B.Re.PM.WVs
Chukar *Alectoris chukar* Introduced
Red-legged Partridge *Alectoris rufa* B.RE. Introduced.
Grey Partridge *Perdix perdix* ScVg
Common Quail *Coturnix coturnix* ScPM Bred prior to 1915.
Common Pheasant *Phasianus colchicus* B.RE. Introduced.
Water Rail *Rallus aquaticus* PM. WVs
Spotted Crake *Porzana porzana* RPM
Baillon's Crake *Porzana pusilla* VVg. Pre 1858.
Corncrake *Crex crex* RPM. Once bred.
Moorhen *Gallinula chloropus* B.Re.WVs
Common Coot *Fulica atra* B.Re.WVs
Crane *Grus grus* VVg
Little Bustard *Tetrax tetrax* VVg. 1853.
Oystercatcher *Haematopus ostralegus* PM.WVs. Has bred.
Avocet *Recurvirostra avosetta* RVg. 1847.
Stone Curlew *Burhinus oedicnemus* RVg
Collared Pratincole *Glareola pratincola*. One found dead 1954.
Black-winged Pratincole *Glareola nordmanni* RVg
Little Ringed Plover *Charadrius dubius* UPM
Great Ringed Plover (Ringed Plover) *Charadrius hiaticula* PM.WVs
Killdeer *Charadrius vociferus* VVg
Dotterel *Charadrius morinellus* ScPM
American Golden Plover *Pluvialis dominica* ScPM
Pacific Golden Plover *Pluvialis fulva* VPM

European Golden Plover *Pluvialis apricaria* PM.WVs
Grey Plover *Pluvialis squatarola* PM
Northern Lapwing *Vanellus vanellus* PM.WVs. Has bred.
Red Knot or Knot *Calidris canutus* UPM
Sanderling *Calidris alba* PM.WVs
Little Stint *Calidris minuta* ScPM
Temminck's Stint *Calidris temminckii* RPM
Least Sandpiper *Calidris minutilla* VVg
White-rumped Sandpiper *Calidris fuscicollis* VVg
Pectoral Sandpiper *Calidris melanotos* RPM
Curlew Sandpiper *Calidris ferruginea* UPM
Purple Sandpiper *Calidris maritima* PM.WVs
Dunlin *Calidris alpina* PM.WVs
Buff-breasted Sandpiper *Tryngites subruficollis* ScVg
Ruff *Philomachus pugnax* ScPM and WVs
Jack Snipe *Lymnocryptes minimus* UPM and WVs
Common Snipe *Gallinago gallinago* PM.WVs
Great Snipe *Gallinago media* VVg
Long-billed Dowitcher *Limnodromus scolopaceus* RVg
Woodcock *Scolopax rusticola* PM.WVs
Black-tailed Godwit *Limosa limosa* PM
Bar-tailed Godwit *Limosa lapponica* PM
Whimbrel *Numenius phaeopus* PM
Eurasian Curlew *Numenius arquata* PM.WVs. Has bred.
Upland Sandpiper *Bartramia longicauda* VVg
Spotted Redshank *Tringa erythropus* UPM
Common Redshank *Tringa totanus* PM
Common Greenshank *Tringa nebularia* PM
Lesser Yellowlegs *Tringa flavipes* RVg
Solitary Sandpiper *Tringa solitaria* VVg
Green Sandpiper *Tringa ochropus* PM
Wood Sandpiper *Tringa glareola* UPM
Common Sandpiper *Actitis hypoleucos* PM.WVs
Spotted Sandpiper *Actitis macularia* VVg
Turnstone *Arenaria interpres* PM.WVs.SVs
Wilson's Phalarope *Phalaropus tricolor* RVg
Red-necked Phalarope *Phalaropus lobatus* RVg
Grey Phalarope *Phalaropus fulicarius* UPM
Pomarine Skua *Stercorarius pomarinus* UPM
Arctic Skua *Stercorarius parasiticus* PM

Appendix: BIRDS 181

Long-tailed Skua *Stercorarius longicaudus* RPM
Great Skua *Catharacta skua* PM
Mediterranean Gull *Larus melanocephalus* UPM and WVs
Laughing Gull *Larus atricilla* VVg
Little Gull *Larus minutus* UPM and WVs
Sabine's Gull *Larus sabini* ScPM
Bonaparte's Gull *Larus philadelphia* RVg
Black-headed Gull *Larus ridibundus* PM.WVs
Ring-billed Gull *Larus delawarensis* RPM and WVs
Common Gull *Larus canus* PM.WVs and occasional SVs
Lesser Black-backed Gull *Larus fuscus* Re.PM.WVs. Breeds occasionally.
Herring Gull *Larus argentatus* B.Re.PM.WVs
Western Yellow-legged Gull *Larus michahellis* RPM and WVs
Iceland Gull *Larus glaucoides* ScPM and WVs
Glaucous Gull *Larus hyperboreus* UPM and WVs
Great Black-backed Gull *Larus marinus* B.RE.PM.WVs
Ross's Gull *Rhodostethia rosea* VVg
Kittiwake *Rissa tridactyla* B.PM.WVs
Ivory Gull *Pagophila eburnea* VVg. 1847.
Gull-billed Tern *Sterna nilotica* VVg
Royal Tern *Sterna maxima* VVg
Sandwich Tern *Sterna sandvicensis* PM
Roseate Tern *Sterna dougallii* ScPM
Common Tern *Sterna hirundo* PM
Arctic Tern *Sterna paradisaea* PM
Forster's Tern *Sterna forsteri* VVg
Bridled Tern *Sterna anaethetus* VVg
Little Tern *Sterna albifrons* UPM
Whiskered Tern *Chlidonias hybridus* RVg
Black Tern *Chlidonias niger* UPM
White-winged Black Tern *Chlidonias leucopterus* RVg
Guillemot *Uria aalge* PM.WVs
Razorbill *Alca torda* B.Re.PM.WVs
Little Auk *Alle alle* ScPM
Black Guillemot *Cepphus grylle* RPM and WVs
Puffin *Fratercula arctica* ScPM. Formerly bred.
Pallas's Sandgrouse *Syrrhaptes paradoxus* VVg. 1888.
Rock Dove/Feral Pigeon *Columba livia* B.Re. It is unlikely that pure Rock Doves exist today in the peninsula.
Stock Dove *Columba oenas* B?.Re. Has bred but no recent records.

Wood Pigeon *Columba palumbus* B.Re.PM.WVs
Collared Dove *Streptopelia decaocto* B.Re.PM. First bred 1962 .
Turtle Dove *Streptopelia turtur* ScPM
Oriental Turtle Dove *Streptopelia orientalis* RPM
Ring-necked Parakeet *Psittacula krameri* Escape.
Great Spotted Cuckoo *Clamator glandarius* VVg
Cuckoo *Cuculus canorus* B.SRe.PM
Yellow-billed Cuckoo *Coccyzus americanus* VVg
Barn Owl *Tyto alba* B.Re
Hawk Owl *Surnia ulula* VVg. 1966
Little Owl *Athene noctua* B? Introduced. Reached Land's End in 1923. Breeds sporadically.
Tawny Owl *Strix aluco* B.Re
Long-eared Owl *Asio otus* ScPM.
Short-eared Owl *Asio flammeus* UPM and WVs
European Nightjar *Caprimulgus europaeus* B.SRe.RPM
Chimney Swift *Chaetura pelagica* VVg
Common Swift *Apus apus* B.Re.PM
Pallid Swift *Apus pallidus* VVg
Alpine Swift *Apus melba* RVg
Little Swift *Apus affinis* VVg
Common Kingfisher *Alcedo atthis* B.Re.PM
Belted Kingfisher *Ceryle alcyon* VVg
European Bee-eater *Merops apiaster* RVg
European Roller *Coracias garrulus* VVg
Hoopoe *Upupa epops* ScPM
Wryneck *Jynx torquilla* ScPM
Green Woodpecker *Picus viridus* B.Re. Colonized about 1870.
Great-spotted Woodpecker *Dendrocopos major* B.Re.ScPM
Lesser-spotted Woodpecker *Dendrocopos minor* RVg
Short-toed Lark *Calandrella brachydactyla* RVg
Woodlark *Lullula arborea* ScPm
Skylark *Alauda arvensis* B.Re.PM.WVs
Horned Lark. (Shore Lark) *Eremophila alpestris* RPM
Sand Martin *Riparia riparia* PM. Nested pre-1900 at Newlyn.
Barn Swallow *Hirundo rustica* B.SRe.PM
Red-rumped Swallow *Hirundo daurica* RVg
House Martin *Delichon urbica* B.SRe.PM
Richard's Pipit *Anthus novaseelandiae* RPM
Blyth's Pipit *Anthus godlewskii* VVg

Appendix: BIRDS 183

Tawny Pipit *Anthus campestris* RPM
Olive-backed Pipit *Anthus hodgsoni* RVg
Tree Pipit *Anthus trivialis* PM
Pechora Pipit *Anthus gustavi* VVg
Meadow Pipit *Anthus pratensis* B.Re.PM
Red-throated Pipit *Anthus cervinus* VVg
Rock Pipit *Anthus petrosus* B.Re
Water Pipit *Anthus spinoletta* PM
Yellow Wagtail *Motacilla flava* PM
'Blue-headed Wagtail' *Motacilla flava flava* UPM
'Ashy-headed Wagtail' *Motacilla flava cinereocapilla* RPM
'Grey-headed Wagtail' *Motacilla flava thunbergi* RPM
'Sykes's Wagtail' *Motacilla flava beema* VVg
Citrine Wagtail *Motacilla citreola* VVg
Grey Wagtail *Motacilla cinerea* B.Re
Pied Wagtail *Motacilla alba* B.Re.PM.WVs
'White Wagtail' *Motacilla alba alba* PM
Bohemian Waxwing (Waxwing) *Bombycilla garrulus* ScPM and WVs. Occasional eruptions.
Dipper *Cinclus cinclus* B.Re.WVs
Wren *Troglodytes troglodytes* B.Re
Hedge Accentor or Dunnock *Prunella modularis* B.Re.PM
Robin *Erithacus rubecula* B.Re.PM
Nightingale *Luscinia megarhynchos* RPM
Bluethroat *Luscinia svecica* RPM
Black Redstart *Phoenicurus ochruros* PM.WVs
Redstart *Phoenicurus phoenicurus* PM
Whinchat *Saxicola rubetra* PM
Stonechat *Saxicola torquata* B.Re.ScPM
'Siberian Stonechat' *Saxicola torquata stejnegeri/maura* RVg
Northern Wheatear *Oenanthe oenanthe* B.PM.
'Greenland Wheatear' *Oenanthe oenanthe leucorrhoa* PM
Black-eared Wheatear *Oenanthe hispanica* VVg
Desert Wheatear *Oenanthe deserti* VVg
Blue Rock Thrush *Monticola solitarius* VVg
Varied Thrush *Zoothera naevia* VVg
Swainson's Thrush *Catharus ustulatus* VVg
Grey-cheeked Thrush *Catharus minimus* VVg
Veery *Catharus fuscescens* VVg
Ring Ouzel *Turdus torquatus* ScPM

Blackbird *Turdus merula* B.Re.PM.WVs
Fieldfare *Turdus pilaris* PM.WVs
Song Thrush *Turdus philomelos* B.Re.PM.WVs
Redwing *Turdus iliacus* PM.WVs
Mistle Thrush *Turdus viscivorus* B.Re.WVs
Cetti's Warbler *Cettia cetti* B.Re.UPM
Grasshopper Warbler *Locustella naevia* B.SRe.UPM.
Savi's Warbler *Locustella luscinioides* RVg
Aquatic Warbler *Acrocephalus paludicola* RPM
Sedge Warbler *Acrocephalus schoenobaenus* B.SRe.PM
Marsh Warbler *Acrocephalus palustris* ScVg
Reed Warbler *Acrocephalus scirpaceus* B.SRe.PM
Paddyfield Warbler *Acrocephalus agricola* VVg
Great Reed Warbler *Acrocephalus arundinaceus* VVg
Booted Warbler *Hippolais caligata* VVg
Olivaceous Warbler *Hippolais pallida* VVg
Icterine Warbler *Hippolais icterina* ScPM
Melodious Warbler *Hippolais polyglotta* ScPM
Dartford Warbler *Sylvia undata* RPM. Bred in 1800s.
Subalpine Warbler *Sylvia cantillans* RVg
Orphean Warbler *Sylvia hortensis* VVg
Barred Warbler *Sylvia nisoria* RPM
Lesser Whitethroat *Sylvia curruca* UPM
Common Whitethroat *Sylvia communis* B.SRe.PM
Garden Warbler *Sylvia borin* B.SRe.PM
Blackcap *Sylvia atricapilla* B.SRe.PM. A few birds overwinter.
Greenish Warbler *Phylloscopus trochiloides* VVg
Arctic Warbler *Phylloscopus borealis* VVg
Pallas's Leaf Warbler *Phylloscopus proregulus* VVg
Yellow-browed Warbler *Phylloscopus inornatus* RPM
Radde's Warbler *Phylloscopus schwarzi* VVg
Dusky Warbler *Phylloscopus fuscatus* VVg
Bonelli's Warbler *Phylloscopus bonelli* VVg
Wood Warbler *Phylloscopus sibilatrix* ScPM
Chiffchaff *Phylloscopus collybita* B.SRe.PM. Some overwinter.
Willow Warbler *Phylloscopus trochilus* B.SRe.PM
Goldcrest *Regulus regulus* B.Re.PM
Firecrest *Regulus ignicapillus* UPM and WVs
Spotted Flycatcher *Muscicapa striata* B.SRe.PM
Red-breasted Flycatcher *Ficedula parva* ScPM

Appendix: BIRDS

Pied Flycatcher *Ficedula hypoleuca* PM
Bearded Tit *Panurus biarmicus* ScPm and WVs. Subject to influxes e.g. 1965 and 1972.
Long-tailed Tit *Aegithalos caudatus* B.Re.PM
Marsh Tit *Parus palustris* Breeds sporadically. RPM
Coal Tit *Parus ater* B.Re.PM
Blue Tit *Parus caeruleus* B.Re.PM
Great Tit *Parus major* B.Re.PM
European Nuthatch *Sitta europaea* B.Re
Eurasian Treecreeper *Certhia familiaris* B.Re.RPM
Penduline Tit *Remiz pendulinus* RVg
Golden Oriole *Oriolus oriolus* ScPM
Isabelline Shrike *Lanius isabellinus* RVg
Red-backed Shrike *Lanius collurio* ScPM. Bred before 1851.
Great Grey Shrike *Lanius excubitor* ScPM
'Steppe Grey Shrike' *Lanius excubitor pallirostris* VVg
Woodchat Shrike *Lanius senator* ScVg
Eurasian Jay *Garrulus glandarius* B.Re.PM
Magpie *Pica pica* B.Re.PM
Nutcracker *Nucifraga caryocatactes* VVg
Red-billed Chough *Pyrrhocorax pyrrhocorax* RVg. Last recorded breeding on the peninsula in 1870.
Eurasian Jackdaw *Corvus monedula* B.Re.PM
Rook *Corvus frugilegus* B.Re.PM
Carrion Crow *Corvus corone* B.Re
Hooded Crow *Corvus corone cornix* ScPM
Common Raven *Corvus corax* B.Re
Common Starling *Sturnus vulgaris* B.Re.PM.WVs
Rose-coloured Starling *Sturnus roseus* RVg
House Sparrow *Passer domesticus* B.Re
Tree Sparrow *Passer montanus* ScPM
Yellow-throated Vireo *Vireo flavifrons* VVg
Red-eyed Vireo *Vireo olivaceus* RVg
Chaffinch *Fringilla coelebs* B.Re.PM.WVs
Brambling *Fringilla montifringilla* ScPM and WVs
European Serin *Serinus serinus* RPM and WVs
Greenfinch *Carduelis chloris* B.Re.PM.WVs
Goldfinch *Carduelis carduelis* B.Re.PM
Siskin *Carduelis spinus* ScPM and WVs
Linnet *Carduelis cannabina* B.Re.PM

Twite *Carduelis flavirostris* RPM
Common Redpoll *Carduelis flammea* PM
'Lesser Redpoll' *Carduelis flammea cabaret* ScPM
Arctic Redpoll *Carduelis hornemanni* RPM
Two-barred Crossbill *Loxia leucoptera* VVg
Crossbill *Loxia curvirostra* ScPM. Occasional irruptions.
Common Rosefinch *Carpodacus erythrinus* VVg
Bullfinch *Pyrrhula pyrrhula* B.Re.ScPM. WVs
Hawfinch *Coccothraustes coccothraustes* RVg
Northern Parula *Parula americana* RVg
Bay-breasted Warbler *Dendroica castanea* VVg
Blackpoll Warbler *Dendroica striata* VVg
American Redstart *Setophaga ruticilla* VVg
Scarlet Tanager *Piranga olivacea* VVg
Lapland Longspur (Lapland Bunting) *Calcarius lapponicus* ScPM
Snow Bunting *Plectrophenax nivialis* UPM.RWVs
Yellowhammer *Emberiza citrinella* B.Re.PM
Cirl Bunting *Emberiza cirlus*
Ortolan Bunting *Emberiza hortulana* UPM
Little Bunting *Emberiza pusilla* RVg
Yellow-breasted Bunting *Emberiza aureola* VVg
Reed Bunting *Emberiza schoeniclus* B.Re.PM.WVs
Red-headed Bunting *Emberiza bruniceps* Probable escape.
Black-headed Bunting *Emberiza melanocephala* VVg
Corn Bunting *Miliaria calandra* RVg. Has bred. Now extinct on the peninsula.
Northern Oriole *Icterus galbula* VVG

Appendix

BUTTERFLIES AND DRAGONFLIES OF THE LAND'S END PENINSULA

These two lists were compiled from the Erica database, from my own records and from consultation with the country recorders for both groups. A Butterfly Atlas for Cornwall edited by A. Spalding and J. Worth (our county recorder) is soon to be published. I have also made references to the following works: The Moths and Butterflies of Cornwall and the Isles of Scilly by F.H.N. Smith (1997) and The Butterflies of Cornwall and the Isles of Scilly by R. Penhallurick (1996). Two older works – The Distribution Maps of the Butterflies of the British Isles edited by John Heath and tetrad square records for the dragonflies of Cornwall drawn up by H. Robinson have also been studied. The status of some species within each group is uncertain because of lack of recording in some parts of the peninsula, as for example with the Marsh Fritillary butterfly. With outside influences (such as lack of grazing, fires etc.) and global warming inducing habitat change, it is important to keep a baseline of information so that population changes can be monitored and new colonizations or extinctions noted. Anyone with contributions should contact the Country Recorders. Descriptions of any rare butterflies or dragonflies are useful, together with grid references and dates.

Nomenclature for butterflies follows the RSNC Guide to the Butterflies of the British Isles by J.A. Thomas (1986) and for dragonflies – the Dragonflies of Europe by R.R. Askew (1988).

Codes for the status of each species are as follows:

C = Common   FC = Fairly Common
Sc = Scarce   Vg = Vagrant
M = Population may be supplemented by immigrants.

Small Skipper *Thymelicus sylvestris* C
Large Skipper *Ochlodes venata* C
Pale Clouded Yellow *Colias hyale* Vg. One record at Sennen, not confirmed, 1991.
Clouded Yellow *Colias croceus* Sc.M. Annual influx of migrants often producing a second brood. Variable numbers each year. Pale female form *helice* often recorded.
Brimstone *Gonepteryx rhamni* Vg. One record 1991.
Large White *Pieris brassicae* C.M
Small White *Pieris rapae* C.M
Green-veined White *Pieris napi* C
Orange Tip *Anthocharis cardamines* FC
Green Hairstreak *Callophrys rubi* Sc
Small Copper *Lycaena phlaeas* C
Small Blue *Cupido minimus* Vg. One record 1983. Unconfirmed.
Large Blue *Maculinia arion*. A colony at Minack Head 1968-69.
Silver-studded Blue *Plebejus argus* Sc
Brown Argus *Aricia agestis* Vg. One Record. Cot Valley 1996.
Common Blue *Polyommatus icarus* C
Holly Blue *Celastrina argiolus* FC
Red Admiral *Vanessa atalanta* C.M
Painted Lady *Cynthia cardui* FC.M
Small Tortoiseshell *Aglais urticae* C
Peacock *Inachis io* C.M
Comma *Polygonia c-album* Sc
Small Pearl-bordered Fritillary *Boloria selene* FC
Pearl-bordered Fritillary *Boloria euphrosyne* R. One recent record of 2 near Carn Gloose 1999.
Dark Green Fritillary *Argynnis aglaja* FC
Silver-washed Fritillary *Argynnis paphia* Vg. Two old records.
Marsh Fritillary *Eurodryas aurinia* Sc. Two known colonies, but may be under-recorded.
Speckled Wood *Pararge aegeria* C
Wall Brown *Lasiommata megera* C
Grayling *Hipparchia semele* FC
Gatekeeper *Pyronia tithonus* C
Meadow Brown *Maniola jurtina* C
Small Heath *Coenonympha pamphilus* C
Ringlet *Aphantopus hyperantus* FC
Monarch *Danaus plexippus* Vg

Dragonflies

Southern Hawker *Aeshna cyanea* Sc
Common Hawker *Aeshna juncea* Sc
Migrant Hawker *Aeshna mixta* Sc
Emperor Dragonfly *Anax imperator* FC
Lesser Emperor Dragonfly *Anax parthenope* RVg Southern Europe.
Green Darner *Anax junius* Vg North America
Golden-ringed Dragonfly *Cordulegaster boltonii* C
Black-tailed Skimmer *Orthetrum cancellatum* Sc
Keeled Skimmer *Orthetrum coerulescens* FC
Broad-bodied Chaser *Libellula depressa* FC
Four-spotted Chaser *Libellula quadrimaculata* FC
Red-veined Darter *Sympetrum fonscolombii* RVg Southern Europe.
Common Darter *Sympetrum striolatum* C
Scarlet Dragonfly *Crocothemis erythraea* RVg. One record.
Beautiful Demoiselle *Calopteryx virgo* FC
Emerald Damselfly *Lestes sponsa* FC
Red-eyed Damselfly *Erythromma najas* R. Pre-1960.
Large Red Damselfly *Pyrrhosoma nymphula* C
Small Red Damselfly *Ceriagrion tenellum* Sc
Blue-tailed Damselfly *Ischnura elegans* C
Scarce Blue-tailed Damselfly *Ischnura pumilio* R. Pre-1960.
Common Blue Damselfly *Enallagma cyathigerum* C
Azure Damselfly *Coenagrion puella* C
Variable Damselfly *Coenagrion pulchellum* R. Pre-1960.
Southern Damselfly *Coenagrion mercuriale* R Pre-1960.

Appendix

MAMMALS OF THE LAND'S END PENINSULA

This list was compiled from my own records and data extracted from the computerised data base ERICA housed at the Cornwall Wildlife Trust, Allet.

Hedgehog *Erinaceus europaeus*
Mole *Talpa europaea*
Common Shrew *Sorex araneus*
Pygmy Shrew *Sorex minutus*
Water Shrew *Neomys fodiens*
Greater Horseshoe Bat *Rhinolophus ferrumequinum*
Lesser Horseshoe Bat *Rhinolophus hipposideros*
Whiskered Bat *Myotis mystacinus*
Brandt's Bat *Myotis brandti*
Natterer's Bat *Myotis natteri*
Serotine Bat *Eptesicus serotinus*
Noctule Bat *Nyctalus noctula*
Long-eared Bat *Plecotus auritus*
Pipistrelle Bat *Pipistrellus pipistrellus*
Rabbit *Oryctolagus caniculus*
Brown Hare *Lepus capensis* Old records
Grey Squirrel *Sciurus carolinensis*
Red Squirrel *Sciurus vulgaris* Last record 1974
Bank Vole *Clethrionomys glareolus*
Field Vole *Microtus agrestis*
Water Vole *Arvicola terrestris*
Wood Mouse *Apodemus sylvaticus*
Harvest Mouse *Micromys minutus*
House Mouse *Mus musculus*
Black Rat *Rattus rattus* Odd records
Brown Rat *Rattus norvegicus*
Feral Goat *Capra sps.* Small herds from time to time.
Red Deer *Cervus elaphus.* One record 2001.

Appendix: MAMMALS

Fox *Vulpes vulpes*
Stoat *Mustela erminea*
Weasel *Mustela nivalis*
Polecat *Mustela putorius* Records from early 1900s. None recent
Mink *Mustela vison*
Badger *Meles meles*
Eurasian Otter *Lutra lutra*
Atlantic Grey Seal *Halichoerus grypus*
Steller's Sea-lion. The one recorded on the Brisons was
  an exceptionally rare vagrant or an escape, probably the latter.

The following list concerns whales and dolphins sighted off the peninsula. The number of entries on the database is given for all except the first four species which are too numerous for extraction. The figures give some idea of the status, although some records will refer to the same animals, for example with the Fin Wales.

Harbour Porpoise *Phocoena phocoena*
Bottlenose Dolphin *Tursiops truncatus*
Common Dolphin *Delpinus delphis*
Risso's Dolphin *Grampus griseus*
Striped Dolphin *Stenella coeruleoalba* 2
White-beaked Dolphin *Lagenorhynchus albirostris*
Orca (Killer Whale) *Orcinus orca* 59
Longfin Pilot Whale *Globicephala melaena* 12
Fin Whale *Balaenoptera physalus* 26
Humpback Whale *Megaptera novaeangliae* 5
Minke Whale *Balaenoptera acutorostrata* 12

The following refers to strandings. A few of these are just outside the area, but they are relevant since they were so close to our waters. Again, the four common species are not listed numerically, as unfortunately strandings are extremely commom due mainly to fishing by-catch.

Harbour Porpoise
Bottlenose Dolphin
Common Dolphin
Risso's Dolphin

Minke Whale 1 (Whitesands Bay)
Orca 1 (Penzance)
Longfin Pilot Whale 50 (east side of Penzance)
Sei Whale *Balaenoptera borealis* 1 (Mousehole)
Sowerby's Beaked Whale *Mesoplodon bidens* 1 (St Ives)
Striped Dolphin 4 (Porth Curno)
White-beaked Dolphin *Lagenorhynchus albirostris* 2 (east side of Penzance and Sennen)
White-sided Dolphin *Lagenorhynchus acutus* 1 (east of St Ives)

Sperm Whales *Physeter macrocephalus* have been stranded elsewhere in Cornwall, as have Fin Whales and Cuvier's Beaked Whale *Ziphius cavirostris*.

REFERENCES

Blamey, M. and Grey-Wilson C. 1989. *The Illustrated Flora of Britain and Northern Europe*. Hodder & Stroughton.

Blight, J.T. 1861. *A Week at the Land's End*. 1973 reprint. Dalwood. Penzance.

Borlase, W. *Natural History of Cornwall*, 1758. New Edition E & W. 1970. Books Ltd. London.

British Geological Survey of Great Britain. 1988. British Geological Survey.

British Geological Survey Sheet 351/358, 1984.

Carew, R. 1602. *Survey of Cornwall*. John Laggard. London.

Corbet, G.B. and H.N. Southern. 1976. *The Handbook of British Mammals*. Blackwell Scientific Publications.

Cornwall Biological Records Unit. ERICA database now housed at the Environmental Records Centre at the Cornwall Wildlife Trust's base at Allet in Truro.

Cornwall Birdwatching and Preservation Society Reports, 1976-1999.

Cornwall Wildlife Trust. Seaquest South-west bulletins.

Dudman, A.A., Richards A.J., 1997. Dandelions of Great Britain and Ireland B.S.B1. London.

Folliot-Stokes, A. *From St Ives to Land's End*. 1908. Greening Co. London.

Folliot Stokes, *From Land's End to the Lizard*. 1909. Greening Co. London.

Forestry Commission Booklet No.20, *Broadleaves*. 1985. Her Majesty's Stationery Office. London.

French, C. Murphy, R. Atkinson, M. *Flora of Cornwall*, 1999. Wheal Seton Press, Camborne, Cornwall.

British Regional Geology. South-West England. British Geological Survey 1975 NERC.

Goode, A.J.J. Taylor, R.T, *Geology of the Country Around Penzance*. 1988. H.M.S.O. London.

Goode, T. Holiday Geology Guides. Land's End and St Ives to Cape Cornwall. 1995. British Geological Survey, NERC.

Hammond, C. 1977. *The Dragonflies of Great Britain and Ireland*, 1977. The Curwen Press, London.

Heath, J. Distribution Maps of the Butterflies of the British Isles, 1982. Unpublished.

Hudson, W.H. 1908. *The Land's End*. Republished in 1981 by Wildwood House Ltd, 3rd Edition.

Hudson, W.H. 1913. *Adventures Among Birds*. Hutchinson Co., London.

Hyde, H.A. Wade, A.E. 1940. *Welsh Ferns.* Revised by Harrison, S.G. 1978. Sixth Edition. National Museum of Wales, Cardiff.

Johns, Rev. C.A. 1899. *Flowers of the Field*, Revised by G.S. Boulger. 29th Edition. The Society for promoting Christian Knowledge.

Lawman, J. 1994. *A Natural History of the Lizard Peninsula.* Institute of Cornish Studies and Dyllansow Truran. Cornwall

Lawman, J. 1997. *Wildlife at Land's End.* Quotes. Northants

Margetts, L.J. Brambles of Cornwall. Botanical Cornwall No 4 19-48, 1990.

Margetts, L.J. David, R.W. *A Review of the Cornish Flora*, 1981. The Institute of Cornish Studies, Redruth, Cornwall.

Margetts, L.J. Spurgin K.L. 1991. *The Cornish Flora Supplement*, 1981-1990. Trendrine Press, Cornwall.

Meteorological Office, *The Climate of Great Britain, the South-West Peninsula and the Channel Islands.* Climatological Memorandum. Revised edition, 1990.

Penhallurick, R.D. 1969. *Birds of the Cornish Coast.* D. Bradford Publications.

Penhallurick, R.D. 1978. *Birds of Cornwall.* Headland Publications, Barton Ltd. Cornwall.

Rackham, O. 1976. *Trees and Woodland in the British Landscape.* J.M. Dent Sons Ltd, London.

Rodd, E.H. 1880. *The Birds of Cornwall and the Scilly Islands.* Trubner Co, London.

Ryves, B.H. 1948. *Bird Life in Cornwall.* Collins, London.

Sharrock, J.T.R. 1976. *The Atlas of Breeding Birds in Britain and Ireland*, 1976. T A.D Poyser. Staffordshire.

Spalding, A. 1992. *Cornwall's Butterfly and Moth Heritage.* Twelveheads Press, Truro, Cornwall.

Spalding, A. *Red Data Book for Cornwall and the Isles of Scilly*, 1997. English Nature, Environment Agency.

Stace, C. 1991. *New Flora of the British Isles.* 1992 Reprint, Press Syndicate of the University of Cambridge.

Stephens, J. *A Cornish Farmer's Diary*, 1847-1918, 1977 Edition. Headland Printers. Cornwall.

Thomas, J.A. 1986. *RSNC Guide to Butterflies of the British Isles.* Newnes Books. London.

Tregarthen, J.C. 1904. *Wildlife at the Land's End.* John Murray, London.

Tregarthen, J.C. 1925. *The Life Story of a Badger.* John Murrey, London.

Vyvyan, C.C. 1948. *Our Cornwall.* London Westaway Books.

Vyvyan, C.C. 1952. *The Old Place.* Museum Press, London.

Watkiss, R. 1978. *Early Photographs of the West Cornwall Peninsula.* Dalwood.

INDEX

Aire Point 145
Alverton 98, 103
aplite 5
apple orchards 103
Armed Knight 34
Baker's Pit 145
Bartinney Downs 145
Basking Shark 48
Birds
    Arctic Tern 37
    Barn Owl 113, 128, 129
    Bewick's Swan 91
    Black Duck 28
    Black Kite 66
    Black Redstart 30, 70
    Black Swan 91
    Black-headed Gull 36, 91
    Blackbird 68, 110, 139
    Blackcap 68, 109
    Blue Rock Thrush 29
    Blue Tit 109
    Bonaparte's Gull 91
    Booted Eagle 66
    Bullfinch 140
    Buzzard 43, 66, 71, 72, 113, 114
    Carrion Crow 41, 114, 131
    Chaffinch 68, 140
    Chiffchaff 30, 40, 109
    Chough 14, 42
    Chukar 130
    Cirl Bunting 127
    Coal Tit 109
    Common Gull 36
    Common Sandpiper 39, 91
    Common Snipe 67. 76, 114, 131
    Common Tern 37
    Coot 88
    Cormorant 37, 91
    Corn Bunting 7, 127
    Corncrake 126
    Cory's Shearwater 32
    Curlew 68, 131
    Dipper 90
    Dunlin 91
    Dunnock 14, 68, 110, 139
    Feral Pigeon 44
    Fieldfare 68, 131
    Firecrest 109
    Fulmar 32, 33, 41
    Gadwall 90
    Gannet 32
    Garden Warbler 109
    Garganey 90
    Glaucous Gull 36, 91
    Goldcrest 108
    Goldfinch 140
    Golden Plover 68, 131
    Goldeneye 90
    Goosander 90
    Grasshopper Warbler 30
    Great Black-backed Gull 36
    Great Shearwater 32
    Great Tit 109
    Great White Egret 86
    Great-spotted Woodpecker 110, 113
    Green Sandpiper 91
    Green Woodpecker 44, 110
    Greenfinch 128, 140
    Greenish Warbler 29
    Greenshank 91
    Grey Heron 85, 86
    Grey Partridge 130
    Grey Phalarope 31
    Grey Wagtail 90
    Guillemot 33, 34
    Gyr Falcon 43
    Hen Harrier 66
    Herring Gull 35, 36

Hobby 66
House Martin 141
House Sparrow 140
Iceland Gull 36, 91
Jack Snipe 67
Jackdaw 11, 41, 114, 131, 142
Kestrel 11, 42, 66
Kingfisher 86, 87
Kittiwake 36
Lapwing 68, 131
Lesser Black-backed Gull 35
Lesser Scaup 91
Lesser Yellowlegs 91
Lesser-spotted Woodpecker 110
Linnet 64, 70
Little Egret 86
Little Grebe 28, 91
Little Owl 43
Long-eared Owl 113
Long-tailed Duck 90
Long-tailed Skua 31
Long-tailed Tit 109
Magpie 130
Mallard 28, 76, 90
Manx Shearwater 32
Marsh Tit 109
Meadow Pipit 38, 41, 64, 68, 69
Mediterranean Gull 91
Merlin 66
Mistle Thrush 110
Montagu's Harrier 66
Moorhen 78, 87, 88, 89, 91
Muscovy Duck 91
Mute Swan 91
Northern Parula 29
Nuthatch 113
Osprey 92
Oystercatcher 38
partridge 114, 130
Pectoral Sandpiper 91
Peregrine Falcon 41, 42, 43, 66
Pheasant 114, 130
Pied Flycatcher 110
Pied-billed Grebe 28

Pintail 90
Pochard 90
Pomarine Skua 31
Puffin 33, 34, 36
Purple Sandpiper 39
Raven 41, 66
Razorbill 33, 34
Red Kite 66
Red-legged Partridge 130
Redshank 91
Redwing 68, 70, 131
Reed Bunting 89
Reed Warbler 90
Ring Ouzel 30, 70
Ringed Plover 38
Ring-necked Duck 28, 91
Robin 68, 110, 139
Rock Dove 44
Rock Pipit 38
Rook 114, 131
Rose-coloured Starling 143
Ruff 91
Sabine's Gull 31
Sand Martin 141
Sandwich Tern 37
Scaup 90
Sedge Warbler 30, 89
Shag 37
Short-eared Owl 66
Shoveler 90
Skylark 40, 41, 68, 69
Snipe 67, 76, 114, 131
Song Thrush 110
Sooty Shearwater 32
Spotted Crake 89
Spotted Flycatcher 30, 110
Spotted Redshank 91
Starling 30, 68, 131, 143
Stock Dove 68
Stonechat 40, 70
Storm Petrel 32
Swainson's Thrush 29
Swallow 129
Swift 30, 142

INDEX

Tawny Owl 113
Teal 90
Treecreeper 113
Tufted Duck 90
Turnstone 39
Veery 29
Water Rail 74, 88
Wheatear 7, 14, 30, 41, 70
Whimbrel 131
Whinchat 30, 70
White-rumped Sandpiper 91
Whitethroat 14, 30, 40, 70
Whooper Swan 91
Wigeon 90
Willow Warbler 30, 40, 68, 74, 109
Wood Sandpiper 91
Woodcock 67, 76, 113, 114
Woodpigeon 68, 115
Wren 68, 85, 110, 139
Yellow-billed Cuckoo 29
Yellow-throated Vireo 29
Yellowhammer 127, 128
Boscawen 101
Bosigran Cliff 14, 31, 146
Boskenna 97, 99, 101, 106, 110, 142
Bostraze Common 6, 77
Bosvenning Common 55, 145
Boswarva Carn 76, 77, 144
Boswens Common 77
Botallack 11
Brewgate Moor 76
Brisons 34, 43, 47
Brown Trout 91
Buryas 3
Buryas Bridge 3, 135
Butterflies
    Clouded Yellow 52
    Comma 7, 132
    Common Blue 54, 72, 117, 121
    Dark Green Fritillary 52
    Gatekeeper 72, 132
    Grayling 53, 72
    Green Hairstreak 73
    Green-veined White 132
    Holly Blue 17
    Large Skipper 53, 54
    Large White 133
    Marsh Fritillary 53, 62, 73
    Meadow Brown 53, 72, 132
    Monarch 52, 95
    Orange Tip 132
    Painted Lady 23, 51, 137
    Peacock 7, 51, 52
    Pearl-bordered Fritillary 52
    Red Admiral 23, 51, 52, 137
    Silver-studded Blue 73
    Small Copper 72, 121
    Small Heath 53, 72
    Small Pearl-bordered Fritillary 52, 53
    Small Skipper 53
    Small Tortoiseshell 23, 51, 137
    Small White 132, 133
    Speckled Wood 117, 133
    Wall Brown 53
Caer Bran 145
Cape Cornwall 2, 22, 24 27, 43
Carn Barges 43
Carn Galver 1, 55, 61, 64, 66, 146
Carn Gloose 48, 52
Carracks 46
Carrick Du 145
Castallack Moor 80
Chapel Carn Brea 145
china clay 6
Chun Down 145
Chyenhal Moor 21, 56, 76, 78, 80
Clodgy Moor 76, 80, 84
copper 6
Cornish hedge 20, 119, 132
Cornwall Wildlife Trust 48, 49, 55, 84, 145
Cot Valley 6, 25, 27, 52, 85, 95, 135, 146
Crean 141
Chysauster 86
Countryside Stewardship Scheme 145
Dragonflies

Azure Damselfly 94
Beautiful Demoiselle 95
Black-tailed Skimmer 93
Blue-tailed Damselfly 94
Broad-bodied Chaser 93
Common Blue Damselfly 94
Common Darter 93
Common Hawker 118
Emerald Damselfly 94
Emperor Dragonfly 93
Golden-ringed Dragonfly 93
Green Darner 95
Keeled Skimmer 93
Large Red Damselfly 95
Migrant Hawker 118
Red-veined Darter 94
Small Red Damselfly 95
Southern Hawker 118
Yellow-winged Darter 94
Drift Reservoir 3, 84, 86, 88, 90
Dutch Elm disease 97, 99
English Nature 145
feldspar 4, 5
Foage Valley 77, 81, 96, 145, 146
furze 55, 56
Geevor Mine 11
glaciation 6
granite 4, 5, 6
greenstone 5
Gurnard's Head 2, 5, 31
Gwennap Head 32, 46, 116
Gwenver 1, 11
Hannibal's Carn 55
head deposits 6
Hellangrove 97, 98
Jubilee Pool 39
kaolin 6
Kemyel Drea 141
Kemyel Wood 15, 97, 101, 104, 145
Kenidjack 11, 15, 20, 25, 84, 85, 95, 146
Kenidjack Carn 74
Kerris Moor 56, 76, 80
Lady Downs 68
Lamorna 10, 15, 85, 90, 95, 96, 97, 99, 103, 110, 114
Land's End 4, 11, 22, 33, 36, 37, 47, 48
Lanyon Quoit 86, 145
Lariggan 3
Levant mine 11
Logan Rock 4, 19
Longships 7
Madron 64, 74, 79, 81, 98
Mammals
    Badger 72, 115, 116
    Bottlenose Dolphin 47
    Brown Hare 72, 114
    Common Dolphin 47
    Common Seal 46
    Fin Whale 48
    Fox 44, 45, 71
    Greater Horseshoe Bat 146
    Grey Seal 46
    Grey Squirrel 100, 116
    Humpback Whale 48
    Killer Whale 48
    Mink 45
    Minke Whale 48
    Otter 45, 77
    Pilot Whale 48
    Porpoise 48
    Rabbit 71, 72, 131
    Risso's Dolphin 47
    Steller's Sea Lion 47
    Stoat 72
    Striped Dolphin 47
    Water Shrew 77
    Weasel 72
Manx Loghtan sheep 14, 146
Men-an-tol 57
Men Scryfa 3
mica 4
Minack Theatre 47, 49
Mining 6, 11, 14
Morrab Gardens 135
Morrab Place 6
Mounts Bay 3, 48
Mousehole 5, 12, 15, 44, 135

INDEX

Mylor slates 5
Nanjizal 17
Nanquidno 6, 11, 83, 95, 147
Nanscawen 101
National Trust 11, 13, 14, 57, 85, 135, 145, 146
Newlyn 5, 36, 84, 98, 103, 115
Newlyn Coombe 85, 99, 106
Newlyn River 3
Newmill 85, 90
Nine Maidens 57
Oak Eggar 52
Paul 142
Peat 56
Pednevounder 19, 37
pegmatite 5
Penberth 10, 13, 15, 37, 81, 84, 85, 88, 90, 101, 104, 110, 135
Pendeen 5, 7, 11, 18, 31, 33, 49, 143
Penlee House Gallery & Museum 147
Penlee Park 135
Penlee Quarry 5, 6
Penzance 103, 139, 140
Penzance Antiquarian Society 147
pillow lavas 5
Plants
    Agapanthus 135
    Alexanders 135
    Ash 58
    Autumn Lady's Tresses 138
    Banana Palm 135
    Bay 103
    Bear's-breech 135
    Beech 98, 99
    Bell Heather 21, 59
    Betony 8, 123
    Bilberry 60, 73, 125
    Birds-foot-trefoil 20, 54, 73, 121, 138
    Biting Stonecrop 136
    Black Spleenwort 15, 61
    Blackthorn 11, 16, 17, 58, 97, 100
    Bladdernut 135
    Blinks 126
    Bluebell 16, 17, 18, 28, 60, 104, 105, 106, 123, 125
    Bog Asphodel 79
    Bog Pimpernel 78
    Bog Pondweed 83
    Bracken 12, 16, 58, 60
    Bramble 16, 127
    Branched Bur-reed 84
    Bristle Bent 60
    Broad Buckler-fern 107
    Brooklime 80
    Buddleia 137
    Bulbous Rush 84
    Bulrush 84
    Butterfly Bush 137
    Carnation Sedge 83
    Chaffweed 126
    Chamomile 126
    Cleavers 17
    Common Dodder 22
    Common Elder 101
    Common Figwort 25
    Common Heather 21, 58, 59, 125
    Common Knapweed 121
    Common Milkwort 22, 59
    Common Reed 25, 80
    Common Sorrel 72
    Common Stork's-bill 26
    Compact Rush 83
    Coral Necklace 126
    Corn Marigold 121
    Corn Spurrey 122, 138
    Cornish Elm 100
    Cornish Heath 21
    Cornish Moneywort 78, 79
    Cornsalad 122, 137
    Cotton-Grass 80
    Cow Parsley 106, 123
    Creeping Cinquefoil 123
    Cross-leaved Heath 21, 59
    Cuckoo Pint 106
    Cuckooflower 81
    Daffodils 7, 12, 13, 104
    Devil's-bit Scabious 62, 73, 138

Dog Rose 100
Dog's Mercury 106
Dog-violet 12, 60, 117
Dorset Heath 21
Dove's-foot Crane's-bill 26
Dyer's Greenweed 22, 73
Early-purple Orchid 106
Elder 16, 58, 97, 101
Eleagnus 99
Elm 16
Enchanter's Nightshade 106
English Elm 99
English Oak 100
English Stonecrop 20, 61, 123, 136
Erica watsonii 21
Escallonia 13, 16, 99
Euonymus 16
Eyebright 138
Field Pansy 122
Field Scabious 125
Fleabane 25, 81
Flowering Cherry 16
Fool's Water-cress 81
Foxglove 8, 19, 20, 59, 123
Fuchsia 16, 101
Germander Speedwell 123
Giant-rhubarb 85
Ginkgo 98
Gladioli 13
Glaucous Sedge 83
Golden Samphire 22
Golden-rod 123
Gorse 8, 11, 16, 17, 58, 73, 97, 100, 125, 127
Greater Bird's-foot-trefoil 25
Greater Stitchwort 123
Greater Tussock-sedge 83
Green Alkanet 136
Ground Ivy 60
Hairy Bird's-foot-trefoil 123
Hard Rush 83
Hard-fern 84, 107
Hart's-tongue 107
Hawbit 138

Hawthorn 100
Hazel 100
Heath Milkwort 22, 59
Heath-spotted Orchid 22, 59, 81
Hedge Bindweed 17
Hemlock Water Dropwort 81
Hemp-agrimony 25
Himalayan Balsam 63
Holly 98
Holm Oak 98
Honeysuckle 17
Hornbeam 99
Horse Chestnut 98
Hottentot Fig 8, 27, 135
Hydrangea 101
Ilex 98
Italian Lords-and-Ladies 106
Italian Rye-grass 120
Ivy 61
Ivy-leaved Bellflower 78, 79
Ivy-leaved Crowfoot 125
Ivy-leaved Speedwell 106
Ivy-leaved Toadflax 137
Japanese Knotweed 85, 135, 146
Judas Tree 135
Kaffir Lily 13
Kidney Vetch 8, 18
Lady's Bedstraw 26
Lady-fern 107
Lanceolate Spleenwort 61
Lesser Celandine 7, 12, 60, 104
Lesser Skullcap 78
Lesser Spearwort 80
Lesser Stitchwort 123
Lime 99
Ling 21
Lords-and-Ladies 106
Lousewort 22, 59
Mahonia 103
Maidenhair Fern 139
Maidenhair Spleenwort 139
Male-fern 107
Marram Grass 26
March Bedstraw 80

INDEX

Marsh Fern 84
Marsh Marigold 84
Marsh Speedwell 80
Marsh St John's Wort 80
Marsh Violet 78
Meadow Vetchling 123
Meadowsweet 80, 81
Medlar 103
Mind-your-own-business 136
Monkeyflower 84
Montbretia 63, 64, 135
Monterey Cypress 15, 98
Monterey Pine 15, 98
Mother-of-thousands 136
Narcissi 13
Nettle-leaved Bellflower 137
Oxeye Daisy 19
Pale Butterwort 77
Pale Flax 121
Pellitory-of-the-wall 138
Pencilled Crane's-bill 136
Persican Speedwell 123
Pink Oxalis 136
Pink Purslane 104
Pittosporum 16, 99
Polypody 61, 108
Poplar 99
Primrose 12, 26, 125
Privet 13, 16
Purple Bugloss 147
Purple Loostrife 25
Purple Moor-grass 60, 77, 79
Ragged Robin 81
Ramsons 106
Red Campion 8, 20, 59, 106, 123
Red Clover 138
Red Valerian 137
Restharrow 26, 123
Rhododendron 64, 101, 147
Rock Samphire 23
Rock Sea Lavender 22
Rock Spurrey 138
Round-leaved Crowfoot 125
Royal Fern 25, 84

Sand Sedge 26
Sand Spurrey 138
Scaly Male-fern 107
Scarlet Pimpernel 122
Scentless Mayweed 121
Scots Pine 98
Scurvygrass 12, 24, 25
Sea Beet 24
Sea Bindweed 26
Sea Campion 8, 12, 18
Sea Holly 26
Sea Sandwort 27
Sea Spleenwort 25
Self-heal 122, 138
Sessile Oak 15, 100
Sheep's Sorrel 72
Sheepsbit 8, 19, 61, 123
Small-leaved Pittosporum 99
Snow-in-summer 136
Snowdrop 103, 104
Soft Rush 77, 83
Solomon's Seal 106
Southern Marsh-orchid 81, 121
Sphagnum Moss 80
Spring Squill 12, 18
Star Sedge 83
Star-of-Bethlehem 106
Sundew 77
Sweet Chestnut 15, 98
Sweet Violet 13
Sycamore 15, 97, 98
Tamarisk 17, 99
Three-cornered Garlic 8, 28, 104, 135
Thrift 8, 12, 18
Toad Rush 84
Tormentil 59
Trailing St John's-wort 123
Tree Mallow 24
Turkey Oak 15, 100
Tutsan 106
Veronica 99
Wall Pennywort 61
Wall Rue 139

Water Figwort 25, 80
Water Hawthorn 83
Water Mint 74, 81
Water Purslane 126
Water-fern 83
Weasel's-snout 123
White Water-lily 83
Wild Arum 106
Wild Carrot 18
Wild Hop 15
Wild Madder 17
Wild Pansy 122
Wild Thyme 20, 61, 138
Willow 11, 16, 76, 97
Wilson's Filmy-fern 62
Winter Heliotrope 136
Wood Anemone 60
Wood Avens 106
Wood Garlic 106
Wood Sage 60
Wood Sanicle 106
Wood Speedwell 106
Wood Sorrel 18, 106, 125
Wych Elm 100
Yarrow 123
Yellow Archangel 135
Yellow Bartsia 121
Yellow Flag Iris 80
Yellow Kowhai 135
Yellow Oxalis 136
Yellow Pimpernel 106
Yellow-rattle 121
Pordennack Point 145
Porth Chapel 1, 10, 21, 27
Porth Curno 1, 10, 17, 19, 85, 141
Portheras 11, 90, 97
Porthguarnon 10, 25
Porthgwarra 4, 10, 12, 17, 20, 22, 25, 27, 42, 43, 44, 84, 130, 145, 146
Porthmeor 5, 77, 96, 126
Porthmoina 84
plate tectonics 4
quartz 5
Rainbow Trout 91

Reptiles
  Adder 50, 51
  Common Lizard 51
  Grass Snake 50
  Sand Lizard 51
  Slow-worm 51
Roskennals 98. 101, 106, 110
Rudd 91
St Aubyn Estate 144
St Buryan 114, 129, 142
St Ives 5, 35, 36, 98, 115, 142
St Just Airfield 138
St Loy 6, 10, 13, 15, 84, 96, 98, 100, 103, 104, 106, 114
Sancreed 114
Scathe Cove 13
Sennen 1, 11, 17, 37, 38, 39, 49, 120, 131
sheep 120
Skewjack 90, 143
Skimmel Wood 97, 103, 110
Stable Hobba 99, 103
swaling 59
Sunfish 49
Tater-du 5, 7, 13, 24
tin 61
Tinners Way 57
Treen 1, 4, 20
Tregothnan Estate 144
Trencrom 100, 145, 146
Trengwainton 97, 98, 101, 113, 134, 135, 144, 145
Trengwaiton Carn 64
Treryn Dinas 19, 22
Trevaylor 97, 98, 99, 101, 113
Treveal Cliff 14, 114, 146
Treveal Valley 96, 101, 141
Trevidder Moor 74, 75, 99
Trewidden Gardens 98, 135
Trungle Moor 76
Turtles 48
Veor Cove 26
Watch Croft 2, 55, 64
Wherry Rocks 3
Woon Gumpus Common 74

INDEX

Xanthoria 5, 139
Zennor 57, 85, 119, 146
Zennor Hill 2, 17

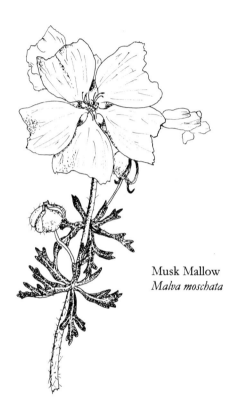

Musk Mallow
Malva moschata